Underground
and Radioactive

Underground and Radioactive

Adventures of a Uranium Miner in 1970s New Mexico

R. D. SAUNDERS

McFarland & Company, Inc., Publishers
Jefferson, North Carolina

LIBRARY OF CONGRESS CATALOGUING-IN-PUBLICATION DATA

Names: Saunders, R. D., 1951– author.
Title: Underground and radioactive : adventures of a uranium miner in 1970s New Mexico / R.D. Saunders.
Description: Jefferson, North Carolina : McFarland & Company, Inc., Publishers, 2017 | Includes index.
Identifiers: LCCN 2017018382 | ISBN 9781476669960 (softcover : acid free paper) ∞
Subjects: LCSH: Saunders, R. D., 1951– | Uranium miners—New Mexico—Biography. | Uranium mines and mining—New Mexico.
Classification: LCC HD8039.U72 U673 2017 | DDC 622/.34932092 [B] —dc23
LC record available at https://lccn.loc.gov/2017018382

BRITISH LIBRARY CATALOGUING DATA ARE AVAILABLE

ISBN (print) 978-1-4766-6996-0
ISBN (ebook) 978-1-4766-2885-1

© 2017 R.D. Saunders. All rights reserved

No part of this book may be reproduced or transmitted in any form or by any means, electronic or mechanical, including photocopying or recording, or by any information storage and retrieval system, without permission in writing from the publisher.

Front cover image of underground uranium mine © 2017 iStock

Printed in the United States of America

McFarland & Company, Inc., Publishers
 Box 611, Jefferson, North Carolina 28640
  www.mcfarlandpub.com

To Marti,
without whom these experiences
would have remained untold

Table of Contents

Preface — 1
Prologue — 6

School Days	7
What Next?	13
The New Hand	24
Camping	33
Orientation	36
The Underground	44
Moving Up	67
Progress	77
Attempted Rape	81
A Place to Call Home	85
Miner's Helper	89
Down the Chute	94
The Iron Blossom	99
Revenge	103
The Sandman	112
A Glimpse of the Future	116
Encino	120
Cal	123
Alone	139

Table of Contents

Darkness	161
Ghost in the Stope	164
Ana Maria	175
Fuzzmobile	182
Beginning of the End	191
End of the Line	194

Aftermath	201
Epilogue	204
Mining Terminology	207
Index	209

Preface

Somewhere in the recesses of my mind has lingered the thought, however farfetched, that maybe, someday, somehow, some way, I could do it all again. That, perchance, the call would go out for experienced though older, much older, former miners. The time has come to acknowledge the personal collection of mining experiences and anecdotes I have been harboring for many years now will be expanding no further, and should be shared.

I suppose for some, who spend the bulk of their working lives moving from one mining boomtown to another, there is nothing especially unique about that life but, unbeknownst to me at the time, my once-in-a-lifetime opportunity was to live and work in one such place in New Mexico, then known as the "Uranium Capital of the World."

Though decades have passed since I last set foot in a uranium mine, not many days go by that I don't miss both working underground and living in a boomtown. It was an exciting time to have lived through and left an indelible imprint on many of us who experienced it.

Although much has been written about the merits and faults of the uranium mining industry, neither of which I touch upon here, frustratingly little has been written about the life of an underground uranium miner.

Even less is known about those working underground at Ambrosia Lake, an ancient, long waterless seabed in northwest New Mexico that is part of the Grants mineral belt.

Over the years I have occasionally told interested listeners about working underground, drilling, blasting, and mucking the sandstone of Ambrosia Lake.

Preface

R.D. Saunders, 1977.

I always enjoyed talking about my time underground, sometimes relishing the incredulity expressed by the listener that often followed the conclusion of the anecdote, but more often appreciating the opportunity to recall the humor of it all because, while mining is deadly serious, much of what happened underground was just plain funny.

The passing of time hasn't tempered the humor much, although it became apparent over the years that my outlook back then was a luxury afforded to me by having no family, no responsibilities, no bills, and no life experience to speak of.

Recounting these experiences took me back underground to the unique sights, sounds, and smells. I especially recalled the characters and personalities that, as the years go by, continue to stand out for their singular uniqueness.

There is no other fragrance or resonance I know equal to that produced by the Jackleg rock drill operating at full bore; no other sight

Preface

New Mexico. The Ambrosia Lake uranium mining district is shown in the upper left.

that matches that of walking up to a miner sitting atop a couple of hundred pounds of dynamite and casually finishing up his cigarette; and no more colorful characters than miners who spent the majority of their working lives underground.

These were the sights and sounds and smells I loved and the people I admired and respected, and each time I had a chance to talk about

Preface

everything and everyone was to once again step on the cage and back underground.

The experiences and anecdotes I share here are true to the best of my recollection. They come from the small world of a sole miner, working on one level of a single mine among over one hundred in the Ambrosia Lake area. My goal is to provide the reader with my interpretation of specific events as I experienced them.

There were many unique personalities to be found underground and unfortunately, due to the confines of the relatively small areas I worked and the limited number of people who were in them, I didn't get to meet everyone at my particular mine, undoubtedly missing some good anecdotes too.

I never worked with a woman during my time underground although I am aware of at least two who worked on another level of my particular mine, so I am unable to comment in this narrative as to what specifically they were doing. However, there were relatively few job classifications in ore production and none of them were easy. If you were there, you did the work.

There were a number of Native Americans, from the nearby Navajo reservation and the Zuni, Acoma and Laguna pueblos, who worked underground but as was the case with women I never had occasion to work with them.

With few exceptions the conversations I recall here are composites of specific situations as I remember them and should not be considered verbatim.

Where possible, I used the actual names of the people I worked with. I infrequently used aliases because while I have forgotten some names, I haven't forgotten the experiences.

A special thank you to the New Mexico Mining Museum in Grants, New Mexico, for their spectacularly authentic recreations of what life was like underground and for preserving the history of uranium mining in New Mexico through their many wonderful exhibits.

The New Mexico Mining Museum is located just off I-40 on historic Route 66.

Please visit the museum's Facebook page for more museum photographs, announcements, and contact information.

Preface

The New Mexico Mining Museum
100 Iron Ave.
Grants, NM 87020

For additional information about Grants, New Mexico, Cibola County, and beautiful western New Mexico, please visit www.grants.org.

Prologue

I thought I was going to live. That hadn't seemed likely an hour earlier when I was fighting to breathe and unable to move. Although I wasn't out of the woods yet, the veil of uncertainty was slowly lifting.

The mine rescue squad had gotten to me quickly, and now I was at the surface again on a stretcher board, waiting for an ambulance to take me into town and to the hospital.

As I lay there staring up at the faces peering down at me, it was the notion that I might get out of this that dulled the searing pain I felt in my back. There had been many injuries in the past two years at Section 35, but this time it was me lying there, looking up at all those faces.

I thought, don't you guys have anything better to do? I was ashamed that a time or two I had been among the circle of faces.

I heard Shotgun's authoritative voice. "Men, we don't need you here; get on out about your work." Thank you, Shotgun.

On occasion I had heard or read about people in situations like I was now in, or worse, where the first thing that came to mind was, how did I end up here? Not so for me. I knew.

The curious crowd dispersed, and I was alone there on the floor, waiting, thinking.

The furthest things from my mind just a couple of years earlier were the thoughts I was having now of how much I'd miss the underground work that I had come to love.

School Days

When I arrived at Illinois Wesleyan University as a new student from suburban Chicago, my first stop had been the front desk of Magill Hall, an all-male residence unit. As I was checking in, the student aide in charge of handing out room keys and other information pertinent to life in the dorm informed me that my roommate would be Cowboy. I looked at the aide and said, "My roommate is Cowboy?"

"Yeah, you got Cowboy. That's your roommate, Cowboy. Room 232. Go up these stairs right here. Matter of fact I think he's up there now."

So, not knowing what to expect, I dragged my footlocker holding all of my belongings up the stairs to the second floor and slowly opened the door to room 232.

There, stretched out on his bed, looking every bit the cowboy with his well-worn boots, jeans, and western-style shirt, was the New Mexican cowboy, Gary Mitchell. I had no way of knowing it at the time, but this was the beginning of my decades-long association with New Mexico.

He seemed to have converted a college dorm room into a kind of bunkhouse. There was a cowboy hat on the dresser and cowboy boots with spurs under the desk.

There was a strange smell in the room unlike anything I had known. Hmm, I wonder what that is, I thought.

But that wasn't the most astonishing thing by a long shot. I noticed Gary seemed to be reading what was clearly a textbook.

The first question that popped into my head had nothing to do with cowboys but was rather, if classes don't start for three days, what's this guy doing reading a textbook? That was Gary, though: usually working at or on something.

Underground and Radioactive

Subsequently, being the gentleman I always found him to be, Cowboy jumped up from his bunk and with an outstretched hand and welcoming smile introduced himself in the slow drawl that is unique to the plains of eastern New Mexico. "Howdy. I'm Gary Mitchell."

Howdy? I hadn't heard howdy much around Chicago or beyond the *Bonanza* television series, for that matter. After the twin initial shocks of seeing him reading a textbook almost a week before classes were to begin and hearing him say howdy in the unfamiliar western drawl had sunk in, I stuck out a sheepish hand and introduced myself.

"So you're the cowboy they told me about," I said.

"Aw ... they call me that 'cause I'm from New Mexico."

"Are you a real cowboy?"

"Yep; family has a ranch in Encino, New Mexico. You don't mind rooming with me?"

"Mind? Why would I mind?"

"Well now, some folks don't much care for the smell."

"OK, now that you mention it, what is that smell? It's not that bad, but what is that?"

"I keep my saddle with me, and some folks don't much like the smell of it."

Looking around the room, I didn't see a saddle, but that peculiar, unfamiliar odor I'd noticed a few minutes before was still lurking about. Still, wanting to get off on the right foot and not finding the smell to be all that objectionable, I answered, "Doesn't smell too bad to me."

"You sure?" he asked.

"Oh yeah; it doesn't bother me a bit."

Our dorm rooms came equipped with a rectangular corner storage unit between the beds about five feet deep, and covered with a heavy hinged lid. Inside our storage unit was Gary's saddle.

While the smell wasn't really that bad, it was certainly sharp and pungent, apparently resulting from a combined mixture of aged, well-worn leather and horse sweat—not something you would expect to find in an Illinois Wesleyan University dorm room.

All the folksy talk could lull people into thinking Gary was some kind of country rube. He was far from it. Gary was quiet, hardworking, studious, and serious. Although I wasn't known for being serious or studious, I was, for the most part, quiet and, if need be, a hard worker.

My dorm room at Illinois Wesleyan. The box between the beds has Gary's saddle in it (photograph by R.D. Saunders).

Gary might well debate how quiet I was, but the pairing worked out well for both of us for quite a spell.

During the first few months of our rooming together, I learned a good deal about New Mexico, a lot about ranching, and a little about horses. During that period occurred the first and only time I ever tagged along on a spur-shopping trip. More importantly I learned how to dress like a cowboy.

Other than in movies, I'd never seen anyone wearing spurs or cowboy boots or even a cowboy hat, for that matter. Then there were the cowboy jeans. Before the era of cowboy-cut jeans came along, I discovered that the way to make jeans more easily fit over boots was to slit them up the side at the bottom hem, fit a piece of triangular fabric between the cut, and then sew it in. There you have boot-cut jeans. Gary's mother made great-looking boot-cut jeans.

Even if I didn't study that much, those things alone were an education that would come in handy later.

I suppose I wasn't taking student life all that seriously, partly

Underground and Radioactive

because I expected to be drafted. The Vietnam War was still going strong, as was the military draft at that time.

During my first year at IWU, I became eligible for the draft. College draft deferments were available but only to those with enough credit hours to qualify. I didn't qualify. I believe I was one hour short of a deferment, so it was a close call.

Thinking maybe IWU could give me a break with that one missing hour, I checked with the registrar's office, but there was no way they could help. I would be classified 1A, a designation given for one year to those eligible for the draft. Those making it through that year without being drafted were free and clear.

Draft numbers were assigned by way of a lottery according to birth date, and my number was 149. I'd noticed that about thirty numbers were being called each month, so it appeared very likely I was going to the army. It didn't happen.

But I was stuck in a self-imposed holding pattern for three or four months just waiting for the letter. I was all packed up and expecting my IWU career to end very shortly. No question, I was not as serious a student as I should have been during that time and was falling behind rather quickly as a result. I soon discovered that might not have been the right approach.

The numbers being called up each month had suddenly dropped off rather dramatically. On December 31, 1971, I ran out to get a newspaper that would have the lottery numbers being called up. Lo and behold, I officially missed being drafted. It was very close, but I was no longer 1A with the draft board.

What I was left with was a lot of catching up to do. That didn't happen either.

I unpacked all my things that had been stuffed into my footlocker and turned to the business of being a college student.

IWU made a point of trying to attract students from all over the United States, and Gary had landed there as part of the effort. The school offered work-study as one component of a financial aid package.

As part of his work-study program, Gary had been employed by the security force patrolling the campus. About the time I was heading to bed, Gary would be out checking and locking doors around campus.

Although I'd had no direct contact with the security force, I'm certain that, through Gary, they knew of me, which came in handy on more than one occasion.

IWU had separate men's and women's dorms at the time and forbade visits to dorms by the opposite gender beyond a certain hour. Whatever hour that was I've forgotten, but it wasn't three o'clock in the morning. That was about the time I was making my way out of Gulick Hall, an all-female residence, one cold winter morning.

Descending the stairs from the second floor of Gulick, I stood inside by a side street door, waiting to make my break. I peered out the window, first left and then right, to make certain there was no security officer nearby. At that hour I was fairly certain the coast was clear, but it was better to be safe than sorry. Whatever the penalty might be for this particular transgression, I didn't want to find out about it firsthand.

Satisfied there was nobody around, I confidently made my way out the door, down the sidewalk, and straight into the path of an IWU security officer. Real nice job checking the coast was clear.

I made a snap decision and attempted to appear as if I was really supposed to be coming out of Gulick Hall at three in the morning. So with as much confidence as I could muster, I walked right up to the security officer, intending to say something, but he simply nodded and passed on by. I couldn't help but think the leniency shown by the officer had been a result of Gary's association with IWU security.

Later, during my senior year, I and a group of other students, with apparently little else to do, organized a student/faculty croquet match. During the lead-up to the match, I had written a story on the upcoming contest for the IWU newspaper to promote the event.

In the completely fictional article, I explained that the match was expected to be hotly contested. I wrote up a fake interview with the IWU chief of security, James Ater. I attributed several quotes to him, including that IWU security would be "ready for anything."

On the day of the match, the participants and spectators were milling around waiting for the games to begin when suddenly there they came—the entire IWU three-man security detail—striding across the quad, decked out in full riot gear, batons in hand, ready for anything. It turned out to be the highlight of the day.

Underground and Radioactive

I ended up rooming with Cowboy for two and a half years, at the end of which I was sad to see him graduate, as that would be the end of our daily association.

But those two years turned out to be the most beneficial education I would get at IWU.

What Next?

By May 1974 my matriculation at Illinois Wesleyan University had come to a close. I probably didn't want it to, though, if my level of preparedness for life after college was any indication.

Only as the commencement ceremony ended did it become clear that the prospect of employment commensurate with my degree was at best questionable—or any employment at all, for that matter. I hadn't exactly been spending all my time looking for work.

Jobless, homeless, and desperate to keep expenses low, I hooked up with another similarly hopeless IWU student, Tom Patterson. In reality Tom's life was far from hopeless, and although he was homeless, his prospects were much greater than mine.

It's interesting how homeless guys gravitate toward one another. Seeing Tom wandering around campus, I was immediately drawn to him and walked up, asking, "You got anything lined up?"

"No, not yet," he said.

It was around two o'clock by then, and not having any plan for that night was not encouraging.

"Got a place to live?" I asked.

"Not yet."

"Me either. Got any ideas?"

"No," he replied, and with that, off we went together, homeless, jobless, into the unknown.

Our predicament called for some creative thinking, so we decided to clandestinely move back into our old dorm, Magill Hall.

Magill was just sitting there empty with open doors and windows, so technically it wasn't as if we were breaking in. It wasn't until years later and multiple tours of college dorms filled with eighteen- to twenty-one-year-old males that the reason for all the open windows at the end of the school year became apparent.

Though Tom and I showed up at Magill only to sleep and used no lights, we hadn't fooled the IWU security force, who, despite our seemingly stealthy ways, appeared before a week had passed and gently suggested we find other accommodations as quickly as possible. That we hadn't been forcefully booted out on the spot was fortunate and probably due, at least in part, to my past association with Gary and the IWU security force.

As it was, Tom and I took advantage of the graciousness shown to us by security by staying put. No, we took our time in leaving Magill Hall and, again, could have done so only through the good graces of IWU security. I have always been thankful to them for that.

Even so, by the end of the second week, we found the doors to Magill Hall closed and the windows locked. That necessitated us having to actually break in just to recover our few belongings.

The prospect of homelessness or of returning to my parents' home was looming ever larger but proved to be a substantial motivational force. Yet it was only through spectacularly good fortune that Tom and I found both a place to live and jobs to pay for it.

Incredibly, I found a job the next day as the new snack shop manager at the first place I visited, the Bloomington Country Club. It was either through a lead that Tom had provided to me or a job that he had held previously, but in any case, I had employment.

The homelessness problem was solved that same day when Tom and I and yet two more wandering former students, Paula Raibley and Marc Brown, combined our resources to land a small furnished apartment. Things were looking up during the summer of 1974.

Snack shop manager at the Bloomington Country Club was a truly wonderful job. The shop, located at the ninth hole of the club, was staffed by me and two coworkers, the brothers Nick and Terry Trang.

Nick and Terry had come to the United States as refugees. The Vietnam War was coming to a close, and they had been lucky enough to get out and had been transplanted to Bloomington.

Both of them were students at Illinois State University in the nearby city of Normal and were just as happy to be working that summer as I was. We were a great crew.

What Next?

Nick was perfectly happy with his Americanized first name, but Terry not so much. I don't know if the brothers were on the opposite sides of the war, but they surely could have been from the sound of things I heard that summer.

The Trang brothers and I came up with innovative ways to keep beer cold, invented several new sandwiches, and waited the better part of the summer for the appearance of *M*A*S*H* star and avid golfer McLean Stevenson.

I was assured by my boss, Mr. Boetcher, that Mr. Stevenson played at least one round of golf a year at the private club and was reminded often of Mr. Stevenson's preference for very cold beer, suggesting I "be ready." Taking that directive to heart, we in the snack shop were always standing by with a full trash can of beer on ice.

When we weren't preoccupied with keeping beer cold, the three us spent many hours experimenting with new sandwich creations, some of which became quite popular with the membership. The beer on ice was a hit too and much preferred over the beer stored in the cooler.

We entertained ourselves by using the snack shop sound system to play tapes of Nick's favorite artist, David Bowie. My guess is that there might be some members of the Bloomington Country Club that remember the summer of 1974 as one of very cold beer and of *The Rise and Fall of Ziggy Stardust and the Spiders from Mars*.

Nick, who, in addition to Vietnamese, spoke several Chinese dialects, was a constant source of interesting information.

While the Vietnam War was winding down and the Kissinger peace plan was dominating the news, I, having been a political science major at IWU, took this unique opportunity to learn from Nick how to curse in Vietnamese and Chinese. Nick was thoroughly helpful in that regard, and the profanities I learned that summer were put to good use over the years.

Terry, on the other hand, was not interested in teaching and remained quiet and rather sullen, probably missing home.

As the summer wore on, it began to look like McLean wasn't going to show up and that our ice and beer in a trashcan idea wasn't going to amount to much other than a spike in beer sales among the regular members. That is, until one day during the final week of the season.

Underground and Radioactive

There came McLean, instantly recognizable, strolling into the snack shop in search of a beer. That was when a summer of hard work paid off. Looking our way, he lifted his can in salute to the three of us with the smiling observation we had worked so hard for: "This is cold beer."

A few days later, the snack shop closed, but having seen it coming and with no desire to again be confronted with the specter of homelessness, I had lined up another job as a graveyard-shift security guard supervisor for Wackenhut Security, which had the contract at State Farm Insurance corporate headquarters in Bloomington.

My roommates having scattered as well that September, I found another apartment, this time alone, and for the next several months, my life revolved around that security job at State Farm and my small second-floor rental unit in the home of a kindly, older widowed woman. I had my houseplants, my books, my stereo, and my job, and that was pretty much the extent of it.

My disposable income from the graveyard shift amounted to just barely subsistence level, but it kept me going. Most days I was sleeping, and during the nights I wasn't working, I tried with some success to alter my circadian rhythm to match the odd hours of the graveyard shift by forcing myself to stay awake listening to the overnight jazz show broadcast on WGN radio out of Chicago.

I read a lot too, sometimes a book a day for a while there. I worked my way through everything by John Steinbeck, Bernard Malamud, Mark Harris, and Herman Melville, to name a few.

Having worked during college for a couple of summers as a security guard, I did have some experience, and I also had that management experience at the country club.

I wouldn't have accepted the supervisory portion of the job, having no desire to supervise, if it hadn't been for the rank that came with the job. Sergeant Saunders sounded good, and anyone who has watched the television series *Combat!* from the 1960s will recognize that name. Hell of a guy.

Then there was the sharp-looking uniform that came with the job, complete with chevron collar devices indicating I was a sergeant.

Yet another advantage was the twenty-five cents an hour pay differential that came with the supervisory position and for working graveyard. When you're looking at a minimum wage of $2.10 an hour, as was the case in 1975, that extra twenty-five cents looks good. Not so attractive was the level of formality.

Bob Carter was the corporate security manager for Wackenhut Security at State Farm, and it was he who had hired me to be his graveyard shift sergeant.

Mr. Carter demanded a higher degree of formality than might have been expected from $2.10-an-hour security guards and insisted that I was to be addressed as Sergeant Saunders and the guards as officer so and so—for example, Officer Jackson or Officer Carcerino.

Although he didn't go so far as to require us to salute each other, most of us, in the spirit of teamwork, mutual respect, and camaraderie, did that anyway. Even the Vietnam veteran on the force, Tom Higgins, was amused and got in on the fun of it from time to time.

For most of the officers, a typical midnight greeting went something like this:

"Good morning, Officer Carcerino."

"Good morning, Sergeant Saunders."

This was followed by salutes and some good-natured chuckles and headshaking among those present for the exchange.

Sergeant Saunders oversaw the nightly activities of seven security officers, who were a mix of college students from Illinois State University in nearby Normal, older adults working a second job, those just barely hanging on to what was clearly a last chance, and the aimless.

As for Officer Higgins, he had been an army medic working helicopter medevac missions during his 1971 tour in Vietnam and was attending ISU on the GI Bill.

Higgins was mostly quiet but sometimes got to talking about his experiences in the war, and after listening to his sobering stories, I was thankful not to have been a medic on a helicopter in Vietnam.

Although he might not have been the most talkative guy, Higgins still did fine work, meaning he always showed up, was awake most of the time, and tolerated the shenanigans of Sergeant Saunders quite well.

Fortunately, for those officers who were students, most of the posts around the State Farm complex were stationary, which meant those officers had no patrol duties and merely sat watching a specific area, such as a loading dock or computer center. Stationary duty stations made it possible to complete homework assignments or simply sit and read. Unfortunately, the stationary posts also made it likely, in fact probable, that certain guards would doze off.

There weren't many things a guard could do to be fired from the Wackenhut State Farm security detail, but sleeping was one of them, and Carter had made it clear that any officer caught sleeping would be dismissed. That was something I had no intent whatsoever of enforcing and little concern of ever having to. After all, it's graveyard shift, I'm the boss, and who but me would know?

Having been a security guard during many, many graveyard nights during past summers, I knew what the job was all about, and it was mostly about nothing. Sit, watch, walk around, wait for the shift to end. If a dangerous situation were to arise, then be sure to leave as quickly as possible. For me it was difficult to fall asleep, but if I had for a few minutes here and there, it wouldn't have mattered.

At State Farm there would always be one guard who roved the complex, stopping at several dozen designated stations along the way as well as all of the stationary guard posts.

The rover was a coveted assignment because it was something to do other than sitting all night and provided some interaction with the few State Farm employees working in the data processing unit. It also provided access to the State Farm employee cafeteria, which, while not open, usually had some unlocked coolers, cabinets, and drawers where goodies might be found. That's security for you.

At the conclusion of each shift, I was required to submit to Mr. Carter my nightly logbook and to provide an oral briefing on the graveyard shift activities. This too was surprisingly formal.

Having been up all night, my goal was to get the briefing over with as soon as possible, so I did in fact make it brief.

The truth was of course that other than all the saluting and formalities, nothing ever did happen worth noting, but the simplicity of my reports never seemed to please Mr. Carter. My mistake.

As time went by, Mr. Carter seemed increasingly concerned about the brevity of my reports and for some reason began asking if anyone was sleeping on duty. "Now, Sergeant Saunders, if anyone is caught sleeping, they're gone," he'd say.

What Next?

Well, of course we had officers sleeping, but I would answer, "Nope, not that I've seen."

"Are you sure? We can't have anyone sleeping on duty."

"No, no, we have a good group of officers, Mr. Carter. No problems there."

As time went on, I sensed a growing desire on Mr. Carter's part to impress upon upper management his overall attention to detail. Perhaps he was looking for a raise or a promotion, or perhaps he felt that if nothing ever happened, then why did State Farm need a security detail? He might very well have needed to justify his existence as head of security and ours as security guards.

I could understand that well enough, so without his saying anything more, I began making an extra effort to find unlocked doors, broken windows, a leaky ceiling, burned-out lights, or anything that could be reported, and I added those to my morning briefings. It was not easy.

Those folks at State Farm ran a nice, clean outfit and didn't leave much around that was broken or left open. Sure, I'd find some cabinets open in the cafeteria, but no way was that getting reported. But other than that, the place was very secure for a building as large as it was.

My minimalist reports did, in fact, do nothing to assuage Mr. Carter's growing apprehension, so he began a series of unannounced visits to the graveyard shift.

Carter would show up, usually between two and three in the morning, looking rather drawn and always serious. These visits would most often consist of a greeting, "Good morning, Sergeant Saunders," followed by a request for a report on how the shift was going, followed by a look at the logbook and a request for confirmation that no guard had been observed sleeping. My report would almost always go something like this:

"Good morning, Mr. Carter. All the officers are at their posts. Officer Carcerino is currently on roving patrol, and there have been no incidents." I would then show him the logbook. With that and a parting "Very well; good night, Sergeant Saunders," Mr. Carter would turn around and be out the front doors and presumably back to bed. It all took no longer than five minutes.

The first time Carter pulled one of these surprise visits, I was just lucky to have been at my desk and actually taking a report from Officer Higgins, who had just come back from a roving patrol. It looked good—perfect, in fact.

The unannounced visits might very well have soothed Mr. Carter's apprehension and given him something to report to his superiors, but they were, by degrees, more and more distressing to Sergeant Saunders.

Knowing there were one or two guards who liked to nap at their posts now and then wasn't comforting, and if by chance Mr. Carter chose to make the rounds of the guard stations in the complex, that would certainly be a problem. The fact that he hadn't up till then lulled me somewhat into contentedness. Could it be, after all, that Mr. Carter didn't want to know about a sleeping security guard any more than I did? Or could it be that he was giving me fair warning? Of course it turned out he was giving me fair warning.

Much like the roving guard, one of my duties was to wander the complex, visiting the various stationary posts, checking on the security officers under my charge, checking the cafeteria, and strolling through data processing looking for unlocked doors.

Finding an officer asleep on these patrols, I would sometimes wake him up, or, hearing my approach, he might wake up on his own, but in any case, the officers knew that I knew they were sleeping. They also knew that I was unwilling to admonish them, or chose not to, let alone take any disciplinary action.

Despite my adding unlocked doors to my morning reports, Mr. Carter continued to suspect my officers of sleeping, as he'd say to me daily, "Now, you're sure nobody is sleeping on post, Sergeant? We can't have anyone sleeping on post."

Finally, during one morning briefing, I was advised that he would be making an inspection during which the both of us would be making the rounds of all the stationary posts. Nice to know.

I immediately informed my staff of officers that the security manager would be making special surprise inspections and that they should be awake (if not alert) at all times for a while.

Sure enough, it was only a couple of days later that Mr. Carter came striding through the front doors at three in the morning.

"Good morning, Sergeant Saunders. Anything to report?"

What Next?

"Good morning, Mr. Carter. Nothing to report. All posts are manned, and Officer Carcerino is on roving patrol."

"Good. Have you made the rounds yet this morning?"

"Not yet."

"Good. Sergeant Saunders, let's make the rounds and check on all the stationary posts."

"You mean together, Mr. Carter?"

"Yes, together, Sergeant."

I was thinking, uh oh. I said, "OK."

I hoped for the best, but knowing my officers, I expected the worst as we left on patrol, figuring Carter would be firing some people that night.

Our first stop was shipping and receiving, manned by Officer Jackson, who was a likeable, very soft-spoken Illinois State student from Chicago. When he was awake, he enjoyed describing his studies and home life.

But Officer Jackson was the sleepiest of officers, so it wasn't a question of *if* Officer Jackson would be asleep but rather how deep a sleep. With luck and a little noise from me, he might hear us coming.

When approaching a stationary guard post, I'd usually jingle my keys, walk heavily, or slam doors in an attempt to wake or warn the guard on duty that I, or someone, was in the vicinity. Often the officer on duty would hear me approaching his post and wake up.

As Mr. Carter and I were making our way down the hallway toward shipping and receiving, I did my best to be as noisy as possible without being too blatant about it, but much to my dismay, as we turned the corner at the loading dock, there was Officer Jackson, book open, pen in hand, but sound asleep at his desk.

Mr. Carter, noting the sleeping Officer Jackson, seemed almost relieved by this discovery and radiated an instant aura of self-satisfaction. It appeared his three o'clock trip to headquarters had indeed been worthwhile. Now, finally, there would be something to report to upper management. We were doing something all right, but it was only catching ourselves not doing our job. Not much for Sergeant Saunders to celebrate here.

There was plenty to eulogize, though, because this was the occasion of the end of the age of innocence for the old Sarge, and in just those few seconds, I knew it.

Underground and Radioactive

The gentle comfort of irresponsibility had begun its inexorable erosion, to be methodically replaced, however slowly, by the encumbrance of maturity. I was downcast but with a little anger mixed in as well. I hadn't asked to be a supervisor, and I wasn't working graveyard shift to actually *do* anything anyway.

Security guards who work graveyard shift at $2.10 an hour at more or less deserted corporate headquarters don't expect to do much of anything. After all, that's why they took the job. That's why *I* took the job. Yet now, here I was, faced with the knowledge that I would be expected to supervise in some way, possibly even wielding authority. I hoped that whatever the discipline meted out to Officer Jackson was to be, it would be done by Mr. Carter in the here and now. It wasn't to be.

Mr. Carter woke Officer Jackson all right, so at least I didn't have to do that. But rather than admonishing Officer Jackson, or Sergeant Saunders, for that matter, he chose to walk off with me in tow, simply saying to me, "Let's go."

We hadn't gone far when Mr. Carter turned to me, asking, "Did you know Officer Jackson was sleeping at his post?"

Thinking quickly, I said, "You mean now?"

"I mean all the time. Is Jackson sleeping all the time?"

"Not when I'm around he isn't," I replied. Now this was marginally true, because, as I said, I tried to make a lot of noise so nobody was asleep by the time I reached his post.

"OK, you're going to have to fire Jackson for sleeping on post."

"I am?"

"Yes, of course. We can't have officers sleeping at their posts, can we?"

"I guess not, but Jackson's a good guy, Mr. Carter. A good security officer."

"No, he's not. You're going to have to fire Officer Jackson."

"Right now? No warning or anything?"

"Yes, now. We have to take security seriously."

I doubt I could have appeared as dejected as I felt, and as I stood there quietly, in the next few seconds an internal argument ensued. Do I fire Officer Jackson? I had been allowing him to sleep at his post for months. How can I fire him now? This was against my principles, such as they were, of being understanding of the struggling student just trying to get by.

What about Jackson's apartment and food and tuition? Then it hit me. Wait a minute—what about my apartment and my food? Suddenly it became clear that I would have to go back to the shipping and receiving loading dock and fire Jackson.

Having made my decision, I slowly shuffled my way back to the loading dock where, incredibly, Jackson was again asleep. I thought, Just great—now I have to wake him up and then fire him. So that's what I did.

"Listen, Officer Jackson, sorry, but Mr. Carter says I have to let you go."

"Let me go. I'm fired?"

"Yeah, I guess so."

"But I have tuition and bills."

Like that was what I needed to hear. I just said, "I know. Mr. Carter said I have to let you go."

"I'm fired?"

"Yeah. I'm really sorry."

"Damn."

With that Jackson dejectedly gathered his books and coat and slowly made his way out of the building. It hadn't been a good night for Officer Jackson or for Sergeant Saunders.

I returned to the sergeant's desk, where, perhaps mercifully, Mr. Carter, having found what he had come for and being anxious to get back home and into bed, didn't even want to visit another post.

Nor did he want to fire me. Jackson's sleeping had been enough. Mr. Carter's existence, and mine, for that matter, could now be more comfortably justified to corporate management. This proved we were doing something.

The incident with Jackson was something I didn't enjoy, and I vowed to make some changes and look for something else to do.

There I was: a college graduate working graveyard shift, making $2.35 an hour, with a car on its last legs, and saddled with some responsibility I didn't care for. That's about the time I heard a knock on my door.

The New Hand

After graduating from Wesleyan, a friend of mine, Greg Hornaday, had made his way to Grants, New Mexico, where he'd found a boomtown that proclaimed itself "The Uranium Capital of the World."

When I heard a knock on my door one afternoon, there was Greg.

He had already started working underground with an outfit called Kermac. He worked at a place called Section 35 and already had some fascinating stories to tell of life underground.

The stories he told of the little city of Grants, New Mexico, made me think of the Wild, Wild West, which was appealing, and he was making what sounded to me like great money. I don't recall what it was, but it was a lot more than the $2.35 an hour I was making working for Wackenhut.

He told me about the mine superintendent, Shotgun Buchanan, and how he needed more hands at the mine. I could have a job if I wanted one despite never having worked underground.

The prospect of being a miner sounded appealing, as did being out in the Wild West, having learned something of it from my old roommate Gary Mitchell, and so did the money I could make.

I thought it over for a day or so and decided that was the place for me.

It took me less than a day to pack everything up, tell Mr. Carter I was leaving, get rid of my car (which would never make the thirteen-hundred-mile trip), and prepare for a new adventure.

Never mind that I really had no job, no place to live, no car, and no money. This was too good to pass up.

It was a long trip out to New Mexico in Greg's 1967 Chevrolet sedan. I'm not really sensitive to motion, be it in a boat or an airplane,

The New Hand

My first residence in Grants. I slept on the floor here and was happy to do so. The rooster next door made sure I was up bright and early each day (photograph by R.D. Saunders).

but riding in a car with no shock absorbers was too much even for me. I threw up a couple of times on the trip out, so it wasn't a great start.

Greg had arranged for us both to stay at the house of an acquaintance of his temporarily, so everything was set. For me that meant a sleeping bag on the floor, but Greg had been staying there awhile, so he had better accommodations.

While Greg had assured me that my hiring on at Kermac, a subsidiary of Kerr-McGee, was a done deal, I was still anxious to get out to the mine to take care of whatever formalities there were and get to work. I was especially interested in meeting the mine superintendent whom I'd heard so much about, Shotgun Buchanan. I didn't know how one acquired a nickname like Shotgun, but it sure did sound interesting. Unfortunately, I never found anyone who knew or was willing to tell the real story behind the name.

Underground and Radioactive

Fortunately, Greg was working swing shift, so he was able to take me out to the mine, known simply as Section 35, bright and early Monday morning.

Most of the mines in the Ambrosia Lake area were somewhere between thirty and forty miles northwest of Grants via NM 605 and NM 509. Judging from the sheer volume of traffic, it looked as if there were a lot of workers at the many Ambrosia Lake mining and milling operations, and it seemed everyone was in a real hurry to get to work.

Traffic was mostly bumper to bumper the entire way along the two-lane road and at least seventy-five to eighty miles per hour. I thought, Not all these people can be running late, so if they really want to get to work in a hurry, that must mean their jobs are a lot of fun.

That kind of thinking showed a little naiveté. Nobody out there was in a hurry because their jobs were fun. It was the driving itself that was fun: part speed and part racing. After making the commute awhile, I came to understand that myself. The drive to work and back really was fun.

There were heavy ore trucks whizzing along among the cars and other smaller vehicles at the same speeds, hauling the end product of all the mining that was going on to one of several area mills, that were the initial processing plants for raw uranium ore.

I made a mental note to watch for ore trucks when it came to be my turn to make the commute, because getting hit by one wouldn't have been good. At the same time, I noted that most of the land along either side of the road had no fences, yet there was plenty of livestock, mostly horses and cattle, grazing out in the distance. I wondered if any of the animals ever wandered out into the road, and if so what would happen.

Of course, various critters large and small did wander onto the road, and what happened was also not good. But that was a lesson in the future.

About halfway out to Section 35, the headframes above the mine shafts began to appear, dotting the landscape along either side of the road, some near and some off in the distance, each one indicating the location of a uranium mine. There were a lot of them operated by several different companies, including Western Nuclear, Ranchers Exploration, Anaconda, United Western Minerals, Phillips Petroleum,

The New Hand

The view we had each day during the drive to work along NM 509 at Ambrosia Lake. There wasn't much opportunity to enjoy it though as the traffic was bumper to bumper at 80 mph or more (photograph by Marti Martienssen).

Homestake Mining, and, where I hoped my new home would be, Kermac Nuclear Fuels.

As vehicles kept turning off left and right to the various mines along the way, the traffic lessened until we eventually came to the Section 35 entrance, that was indicated only by a large metal gate.

The gate being open, we turned left onto a very rough dirt road and over a cattle guard. I supposed everyone simply had to memorize the route, because there were no signs indicating which headframe belonged to which mine. That turned out to be true, as I later missed the turn a few times before learning the way.

We stayed on that dirt road for a while, passing a couple of others, before the large headframe of Section 35 came into view, and we pulled into the large graded area that served as a parking lot.

I observed that in addition to the headframe, there was a large, oblong metal building and some very large piles of rock and sand, but not much else was visible.

Underground and Radioactive

Everyone I saw seemed to be either entering the large metal building or standing around by the headframe. Had I paid more attention to the men waiting by the headframe, I would have saved myself a good deal of anxiety later on, but I was focused on meeting Shotgun and getting done whatever it was I needed to do to get hired.

I was a little anxious as we walked toward the entrance of the large metal building. The men I saw looked rough and they looked tough, and at the time I was neither. Many of them knew Greg, though, and there was some back-and-forth banter as we passed. I don't recall now what was said, but I don't think it was about me being a new guy, or I would have been more nervous than I already was. As I learned soon enough, new guys were mostly ignored, and this was my first taste of it.

Walking through the door of that large metal building was entering a strange world, the world of underground mining. I'd estimate the floor was maybe five thousand square feet. Over most of it were rows of long wooden benches. Here and there were some square wire mesh baskets suspended by cables hanging from the ceiling while others were perhaps 5 feet above floor level. Next to the baskets, just above the floor, were a few men who appeared to be changing from street clothes into their work clothes, and as they did so, they hung their street clothes onto hooks attached to the wire baskets.

As most of the men for this shift were already changed and waiting out by the headframe, almost all the baskets with their hanging clothes had been hoisted to the ceiling. It was an odd sight, looking up and seeing all those baskets, but it later became obvious what a necessary and efficient system it was and why, as the place where wet work clothes were hung at the end of a shift, the large metal building was called the Dry.

After entering the Dry, the first door to the left led to the Section 35 offices occupied by various foremen, engineers, geologists, supervisors, and of course the mine superintendent, Shotgun Buchanan.

But first Greg introduced me to the assistant superintendent, Mel Vigil. Now here was a tough-looking older man, who I guessed was possibly in his late forties and who appeared to have seen it all. I didn't know it at the time, but Mel Vigil was a kind of mining legend around the Ambrosia Lake area who had, in fact, seen a lot and had probably done it all as a first-rate miner. Maybe he was tired of the grind, tired of the danger, or just plain tired out all around, but now he was an assistant superintendent who acted as a liaison of sorts between Shotgun and the miners, and although it later struck me that he would rather have been mining, I think he was good at his job.

Whatever my appearance seemed, I know it put a cross between a concerned and bemused look on Mel Vigil's face. I don't think he disliked me, but it was somewhere in between the two, I suppose. I could probably guess what it was for many years and never know. In retrospect, knowing what I do now, and having had the kind of experience Mel did, were a new hire to have appeared in my office looking as I did that first day, I would have had to suppress mightily the urge to burst out laughing.

Nevertheless, perhaps inexplicably, I liked Mel right away as he briefly explained who he was and what he did at Section 35. I do know that as things turned out, he was always fair with me, and I couldn't have asked for any more than that.

Leaving Mel's office, we next knocked on Shotgun's door. It was open anyway, but it seemed the polite thing to do. As Greg introduced me to Shotgun, I didn't notice any bemused look. If first impressions mean anything, I don't think he liked me very much. It could have been that he was in the middle of something or had other things on his mind, but his look said he wasn't really happy to see me right then. I'll never know and it doesn't matter, but it's another thing I've wondered about for many years, in light of events as they ultimately transpired.

His given name Arnold, Shotgun was a tall, wiry, good-looking guy who I'd been told came to Grants from Butte, Montana. I didn't meet many fellow workers with college degrees but Shotgun was one of them, having earned a mining engineering degree from Montana Tech in 1970, after which he went to work for Anaconda Copper Mines. When copper production in Butte began to rapidly decline he had moved to Grants where his expertise in a specific type of mining, that I'll get to later, was put to good use.

I'd been instructed by Greg never to refer to Shotgun as Arnold. I could call him Shotgun, boss, or anything other than Arnold. Well, with a name like Shotgun, why would I call him anything else?

He was a little gruff at first, asking me what made me think I could be a miner, to which I replied something to do with working hard, being on time, and learning fast—none of which mattered anyway because he needed people with or without experience, and I knew it. Maybe that's what annoyed him, the fact that he knew that I knew and it really didn't matter who or what I was. So he said he could use a guy like me on his crew but that I still had to apply at the Kermac office. He told me he'd call over there to let them know I was on the way and that I was to be assigned to Section 35. I said thanks and good-bye and headed over to the Kermac Ambrosia Lake Division office.

The division office wasn't far from the mine, so we were there in less than fifteen minutes. True to his word, something I found out later he always was, Shotgun had already spoken to the personnel manager, who was expecting me and seemed much happier about it than either Mel or Shotgun had been. They just flat out needed people, so Kermac Human Resources was always hoping guys like me would walk through the door.

I was given some forms and directed to a small area in the office with school desks where I found a few others filling out forms. Some of the others were like me, meaning zero mining experience, but some of the men there sure looked different than I did. Even though I had little to go on, they looked as if they had some experience doing this. I was aware that if I was really going to be a miner, then that's what I would eventually look like.

I filled out all the required paperwork and turned it in, expecting maybe a short interview of some kind before being sent back out to Section 35 and heading on underground as an official miner. Well, of course nothing is ever that easy, even when the company you're applying to is desperate for people.

Looking over my completed paperwork, the personnel manager might have asked me a question or two, but the only thing I remember being told was to report to the Grants Clinic for the Kermac physical exam.

That's something I hadn't thought of, but it made sense. So, fine, I'd go for my physical and then report to Section 35 and head on underground as an official miner. My naiveté was again exposed when the HR manager said, "If you pass the physical, report here on Wednesday for orientation."

The New Hand

"Orientation?" I said.

"Right. All new hires go through a three-day orientation course so you'll know how we do things here. You have never worked underground either, so it'll be important."

The entire hiring process was already taking a lot longer than I had expected, but, hiding my disappointment as best I could, I thanked the manager and went back into town and to the Grants Clinic.

Again they also seemed to be expecting me at the clinic. I was given some forms to fill out, which I did, and then was told to get in a line of men already waiting at a check-in desk. I assumed every mining company sent their new hires to the Grants Clinic, which accounted for the long line. I thought, This can't all be Kermac people, can it? Before long I checked in and sat down to wait with the others.

I had to hand it to them, the Grants Clinic had an efficient assembly line set up, so it wasn't long before my name was called.

The first stop was the weigh-in, followed by the standard physical exam. No problems there. Then it was to the hearing test. I wasn't sure why I needed a hearing test, but OK, no problem.

Upon entering a soundproof booth, I was directed to put on a set of headphones and raise my hand every time I heard a tone. I sat there watching the nurse and technician running the test and listening for the tones but heard nothing. It seemed like quite a bit of time had elapsed, but I hadn't heard a thing. The nurse and the tech didn't appear to be waiting for anything, so I figured the test must be working. So, still not hearing a thing, I raised my hand. This seemed to amuse the two of them, and I immediately heard a faint tone. Lowering my hand, I sat there listening and then straining to hear something—anything—but no more tones were coming through, so I raised my hand again and immediately heard another tone, even more distant than the first.

By this time the nurse and tech were about beside themselves without actually bursting out laughing, but I wasn't finding any of it the least bit funny. Apparently giving hearing tests to prospective miners all day, every day, can get monotonous and lead to some creativity, because these two clowns were making a real game of it. In the end I passed the hearing test but wasn't happy with those two.

Somebody then pulled me aside and sent me over to do a blood pressure recheck in which the nurse in charge attached the cuff and took the reading, followed by a "Hmmm, let me take this again." That was followed by another "Hmmm." She then sent me with my test results in hand to see the physician on duty.

The reason for the nurse giving me the "Hmmm" soon became apparent when the doctor asked, "Have you ever been diagnosed with high blood pressure?"

"No, not that I know of."

"It's too high to pass the physical, so we are going to have to get it under control."

That seemed odd to me because I'd never had any blood pressure issue that I knew of. The doctor handed me a bottle of pills, telling me to take one a day and return in a week. If my blood pressure was within the normal range, he would pass me, and I could report for duty at Section 35.

It was crushing news. I had expected to begin work right away, then had found I had to fill out paperwork, then had been sent to the clinic for a physical, and now this. I was moving further away from going underground with every step, but there wasn't a thing I could do except follow orders and start taking the pills. So in bitter disappointment, I headed back to what was now home, the Coal Mine campground in the Cibola National Forest.

Camping

Uranium mining was reaching its peak in 1975, and Grants was teeming with new arrivals, all looking for a place to call home. A few mining companies were providing housing in enclaves of mobile homes for their most essential employees, and the rest of us were on our own to work out whatever living arrangements we could.

Would-be miners by the hundreds all scrambled to find housing but found there was seldom a house available for sale or rent and even fewer apartments. Most places, if they became available, were snapped up sight unseen. So it was fortunate for us that Greg had some friends who agreed to let us stay at their place for a few days.

There were a number of small homes built into the hillside on the northwest side of Grants that housed nomadic young men like us. A narrow, rutted road that had once been asphalt but now alternated between dirt and gravel wound its way along the hillside among a hodgepodge of ramshackle dwellings that had seen their best days back in the 1950s. They were now in an accelerated state of disrepair exacerbated by itinerant tenants like us, but it didn't matter. I was happy to have the floor to sleep on and the roof over our heads.

The first morning, I remember being awakened by a rooster that seemed to be somewhere inside our home but was actually at the house next door. That was a surprise, both because I'd never had a morning alarm quite like that and because while the town seemed small, it was, after all, a town, not a farm.

The distinction between rural and city sometimes got a little hazy in a small town like Grants, so it wasn't unusual for people moving from farms and ranches to bring their animals with them to live in town. The rooster was a popular choice.

After a few days of the five-in-the-morning rooster call, I began to wonder if the murder rate was above average in Grants—with one neighbor killing another over a rooster.

Rather than look for a cramped living space of some kind in the Grants or adjacent town of Milan, our master plan was to live in the National Forest Coal Mine campground located up on Mount Taylor, at least for the summer months. After that we hoped to find a decent place in town, one that was relatively quiet and didn't come with roosters.

We had each brought with us all the camping gear we owned individually, which together amounted to a large cabin tent and a small ax. The local discount store provided us with the rest.

The Cibola National Forest covers a vast area of New Mexico, Oklahoma, and Texas and includes the 11,300-foot Mount Taylor just north of Grants.

Named in 1849 for then president Zachary Taylor, Mount Taylor is the largest volcano of an extinct volcanic field dating back 1.5 million years. Massive lava flows, known as El Malpais, cover thousands of surrounding acres. El Malpais is today a national monument.

To the Navajo, the Acoma, Laguna, Zuni Pueblos, and the Hopi tribe, Mount Taylor is a sacred site. Known as Tsoodził, or turquoise mountain, Mount Taylor marks the southern boundary of traditional Navajo territory. Beneath it lies a rich deposit of uranium-vanadium, the potential mining of which was, and is, a source of contention between the Navajo and the State of New Mexico.

There were several mineral claims made on Mount Taylor and even some exploration by mining companies under the mountain, but no extensive uranium mining was ever carried out, and development was halted when the boom petered out in the 1980s. In 2009 Mount Taylor was placed on the National Trust for Historic Preservation's list of America's Most Endangered Places.

During my time living in the Grants area, I spent many days hiking in the Mount Taylor area. Occasionally I would hike to the peak, where a mining company had placed a claim stake. Having been up there and seen the beauty of it, I am very happy no mining ever took place on the mountain. There were other places and other ways to get the ore.

Mount Taylor and the surrounding area of the Cibola National Forest are beautiful places in the high desert country of western New Mexico. There are some truly grand vistas available from several areas

Camping

on Mount Taylor. Many times I would sit on the edge of Horace Mesa with my feet dangling over the side, just looking out over the vastness below. Gorgeous natural beauty made it very easy to imagine how the area would hold spiritual significance to many native peoples.

Coal Mine campground was in the 1970s and remains today the lone developed campground on Mount Taylor. When Grants was booming, Coal Mine campground could be found full to capacity every night, so it was with some good fortune that Greg and I found an empty space there to call home.

We had a great time living at Coal Mine campground. It was usually cool up there at night, making it wonderful for the sleep that I got a lot of at first.

Coming from sea level to seven thousand feet turned out to be an interesting experience. The relative lack of oxygen caused me to tire easily, and I found myself sleeping at least twelve hours a day. It took me quite a while to figure out what was happening.

At Coal Mine we had a nice, basic setup with a large picnic table, plenty of shade, full-service restrooms, and a large, flat area to pitch a tent. It was usually a peaceful spot, but not always, as we found out a few weeks later.

We commuted from Mount Taylor to Section 35 each day for the first month or so, and that added another ten miles to the daily drive. I got tired of the extra mileage and came up with an alternate route over forest service roads that cut fifteen miles off the commute but added another thirty minutes to the drive. Those were terrible, terrible roads back there—Range Rover maybe, but a 67 Chevy Bel-Air? Hello, repair bills. That experiment ended in a hurry.

The campground was usually full of people looking for work, but I didn't know anyone like us who had taken up residence. There was a twenty-one-day limit on stays at Coal Mine, after which campers were required to leave, but they could come back later in the day, and if there was a vacancy check back in for another twenty-one days. When our first stay was up, we did just that, checking in for another twenty-one days.

The search for housing in Grants, Milan, and beyond was going nowhere but not for lack of trying. There were just very few vacancies, and those that came up went to friends and relatives of the departing occupants without ever being advertised.

Orientation

About the high blood pressure diagnosis, I have never been sure. It didn't seem likely to me, but I started taking the pills as prescribed, hoping against hope I could pass the physical and finally get to work. The following Monday I drove back to the clinic for a recheck of my blood pressure. I had taken those pills each day as instructed.

It was all over in five minutes as I passed the blood pressure test. With my certified physical exam in hand, I went back to the Kermac office, where I was scheduled for a three-day orientation beginning that Wednesday.

Every new Kermac hire was required to attend a company orientation class whether or not they had any mining experience, mill experience, or underground experience. It was a three-day course that explained how things were done at Kermac and what the federal and state rules and regulations were pertaining to underground workers. Orientation concluded, everyone would report to their respective mine.

Not knowing what to expect, I reported to new employee orientation on Wednesday, ready for anything.

As I looked around the room, it occurred to me that I was in a distinct minority. The guys there looked like they had worked in this business before, but more importantly and more confusingly, they were speaking in a vernacular I had never heard before. I knew they were discussing mining, but that was about the extent of it.

I overheard a conversation between a couple of men, who it seemed had worked together before, talking about a third. "Where ya been, pard?" said the first.

"Been workin' in Green River at the Tony M," the other replied (many mines were named after people). "Fuckers don't pay contract."

"Fuck that. Didn't Simpson tramp out there?"

Orientation

"Fuckin' A. Fuckin' missed a cutout. Fuckin' motor got 'em. Hospital. Laid up fer a while now, pard."

"No shit; that'll do it."

I didn't know what a motor was or what a cutout was or why Simpson the tramp was working in a mine, but it didn't sound like whatever had happened was good when I heard them mention a hospital.

Conversations like that were going on all around me when into the room walked another rather grizzled-looking older man, who announced the beginning of Kermac orientation. I have long forgotten the instructor's name but have been able to piece together the gist of what he said to us. His brief opening statement went something like this:

"How many of you men have worked underground?"

As luck would have it, all hands went up except mine. That right there made me stand out, and I can unequivocally assure you that being a new guy who knows nothing about mining isn't the preferred method of standing out in the company of miners.

Our instructor, to whom I was already profoundly grateful for not pointing me out, mercifully ignored me and continued to address the group, or more specifically, me.

"OK, you new guys, listen up. Mining isn't for everyone, and working underground isn't for everyone. Let me tell you right here, if any of you guys get off the cage and can't take it, then you just get right back on and head back up, and nobody will think anything of it."

I wondered what the "cage" was. It didn't sound good, but I figured it must be something that has to do with getting underground. What kind of business was this that a guy could get off the cage and immediately not be able to "take it"?

Can't take it, eh? That did little more than set my resolve. I wasn't going to be the guy who "can't take it." The instructor continued.

"Now, I don't know where any of you men have worked before, and it don't matter, but here at Kerr-McGee we take safety seriously. Safety first, men. You'll always scale the ribs and back before advancing. We don't want nobody working any ballrooms. There won't be any blasting except at lunch and after shift. We don't hand light rounds here; everything is electric with central blasting.

"You always wear your safety glasses and ear plugs. No smoking underground, men. Take your ventilation with you at all times. We test here, and if you get too much radon, you're done and on the surface where you can't make money, so like I said, take your ventilation with you. And all you men know if your pard don't show up, you can't work on yer own. You either get a temporary helper that day or you don't work."

Then, again speaking to me without looking my way, he said, "And you new guys, you laborers, listen up: we never work alone here. Like I said, you got a pard with ya, or ya don't work.

"If you run a motor, you pull every trip light. If you're walking the main drift, be aware of where the cutouts are. We don't want nobody getting run over again. Make sure your battery is charged before shift. Be sure to tag in and out. We don't want to waste time looking for yer ass if you ain't really missing or dead. Oh, and here at Kermac you'll all have a dog tag attached to your belt with your name and social. That way when something happens and there's not that much left, at least we'll know who it was that there's not much left of." How comforting that all was to hear.

Based on the looks the men were giving each other and the commentary running under

Miner's belt with a pipe wrench holder made of an old piece of water hose at the upper left and a dog tag riveted on. If something unfortunate were to befall a miner the dog tag might survive and provide identification. That was the reasoning, anyway (photograph by R.D. Saunders from an exhibit courtesy New Mexico Mining Museum).

Orientation

their breath around the room, Kermac orientation was the industry standard at every mining company and usually consisted of the same basic spiel from a retired miner converted to an instructor. Most of the guys around the room were either disinterested or nonchalant about the whole thing, so I tried my best to look and act the same. That didn't last long.

Orientation went on like that for some time as the instructor emphasized all the areas he would later be going over in detail. Most of the mining terminology whizzed over my head, but from the sound of it, mining could be dangerous, and safety was a good idea. If nothing else, Kerr-McGee orientation class impressed upon me the importance of safety.

Sometimes I'm asked how safe it was to be around uranium ore. I learned in orientation that a small chunk of uranium ore has about as much radiation as a bunch of bananas. Much later I looked that up and found it to be true as bananas contain radioactive isotope potassium-40.

I had many questions, the sheer number of which would have taken up an hour or more. I had no intention of drawing that kind of attention to myself among this group of veteran miners. I wasn't about to ask what a drift was, what a ballroom was, what a trip light was, and what exactly was meant by "run over again." But it seemed safety was to be taken seriously, and I liked the sound of that. More proof I knew nothing about contract mining, where every safety regulation cost a miner time and thus money.

Orientation continued for the promised three days, and as it progressed I slowly learned what many of the terms being referred to meant. But having never been underground, I found it difficult to visualize what a mine really looked like. Still, I resolved not to make any inquiries of the instructor so as not to draw attention to myself as the no-experience new guy.

Today if I am ever asked what a mine looks like, I'll say, "Think of an ant farm." Looking at a map of the mine I worked in always reminds me of an ant farm.

Smoke and coffee breaks came and went, during which there were many opportunities to talk to the real miners, but I didn't try it. They stayed to themselves, and none of them seemed approachable; certainly none bothered to so much as look my way, and I don't blame them. I couldn't speak the language anyway, so even if anyone had shown some empathy, I couldn't have said much other than to ask a slew of elementary questions.

It was in orientation that I discovered I was not a miner, not because I had never been underground before, but rather no matter how far underground I might venture, it seemed the title of miner was something to be earned not bestowed upon just anyone.

Unfortunately for my ego, it became clear during the class that just because I was an employee of Kerr-McGee and would be working underground, I wouldn't be a miner. That was disappointing. Going in I'd simply assumed that everyone working underground was a miner. I'd seen the photographs of the guys with the hard hats with the lights on top and thought that those guys were miners. That evidently was not necessarily the case.

Orientation taught me that I was a laborer who happened to be working underground and wouldn't be doing any mining at all and would instead be performing any task the shift boss deemed necessary, other than mining. In fact, as it turned out, I wouldn't even see a real miner except before shift, at lunch break, and after shift. I was very disappointed.

Although the instructor covered a wide variety of topics, he missed a couple of the important ones that would have been useful to any new guy going underground for the first time. One of them was what to wear.

When I had reported to Shotgun at Section 35, I thought I was dressed for work underground with my straight-leg Levi's, flannel work shirt, and leather roper boots. If I had been allowed underground dressed like that, then whatever work I managed to get accomplished would have quickly shredded my clothing and ruined my boots in a matter of hours.

Fortunately, the instructor did hand out a list of all the equipment each worker would need to purchase before reporting to their mine. I was probably the only person in orientation who took that paper with me, because those other guys already had their gear.

There was a lot of personal equipment on the list, including a hard hat, miner's belt, miner's rubber boots with steel toes, pipe wrench,

Orientation

safety glasses, gloves, and ear plugs. Sadly, there was no mention of what type of clothing a laborer should wear.

There were several mine supply stores in the Ambrosia Lake area where all the necessary items mentioned on my list could be purchased. My biggest worry was that I needed size 14 steel-toed, high-rise rubber boots. Shoes could always be a problem in general, I'd found, but to the mining supply companies in the area, it wasn't. Lots of large men worked the mines, so my size 14 boots were readily available. The bigger issue was the hard hat.

There are a few shapes of hard hats that underground workers wear. Selection is by choice, and while style is a part of it, functionality is a larger component. Those who work beneath the water table in predominantly wet conditions tend to wear hard hats that have a wide brim, while those working in a dry area of the mine often use a hard hat that more closely resembles a baseball batting helmet.

There were various colors of hard hats to choose from too, but other than style preference, I never knew the significance of them. Although I preferred the baseball type, in the end I chose a hardhat with a brim that went all the way around. If I worked in a wet area, I'd be covered, and if I worked in a dry place, the hat would still work. My biggest problem in the hard hat area was that it was new and shiny white. I might as well have worn a nameplate on my head that said, "New Guy."

It was easy to identify old hands by their hard hats. Most guys tended to advertise on them all the companies they had worked for during their careers by affixing company stickers. But the main feature of the hard hat of an experienced hand was that it was beat up. Anyone who had worked underground for any length of time had a hard hat that was pretty well beaten and scraped and permanently dirty, while a new guy had a hard hat that was clean and shiny, sticking out like a gold nugget in a sea of coal. That was to be me with my new, pristine, pretty white hard hat.

There are a couple of other pieces of mandatory equipment I should mention. One is the pipe wrench. Every person who works underground has one, as its uses are many. It can be anything from a hammer to an actual wrench, but you quickly learn it can be used in many ways. I would never leave the Dry without it attached to my safety

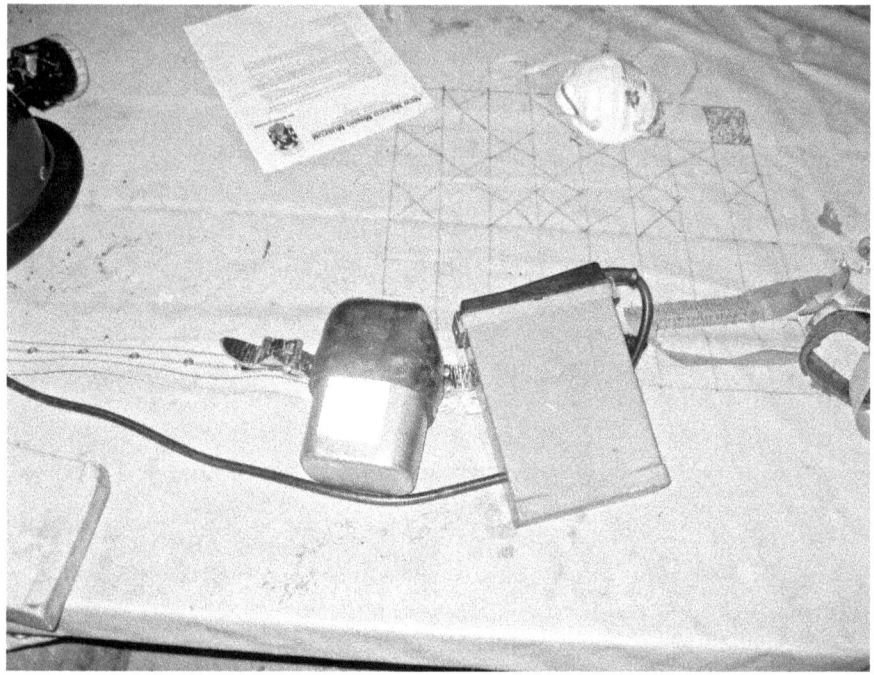

Self-rescue unit (left) and lamp battery (right). Worn on the miner's belt, they were a heavy combination. The self-rescue unit provided emergency short-term air purification against toxic carbon monoxide such as that encountered by a miner caught in smoke or fire (photograph by R.D. Saunders from an exhibit courtesy New Mexico Mining Museum).

belt in its homemade pipe wrench holster made of a piece of old rubber hose.

The other piece of mandatory equipment was the self-rescue unit. Worn on the belt, the self-rescue unit supposedly provided a worker with about fifteen minutes of breathable air should he find himself in a situation where it was needed. It would be a while before I found out what that situation was, but carrying it did seem like a good idea. So despite its weight, I happily lugged it around.

Besides being heavy, the main drawback to the unit was that it became extremely hot during use, and the wearer could suffer burns to the mouth. I found another inconvenient aspect of the self-rescue unit to be that it occasionally made it difficult to squeeze through tight spaces, of which there were many underground, and that it frequently got caught on ladders in manways.

I knew the work was rough, so I had decided to wear the most durable clothing that I had, my straight-leg Levis and a flannel work shirt. Not good, but it was all I had and all I knew.

My clothing problems began with the straight-leg pants. The high-topped, brand-new rubber boots didn't fit under the straight-leg jeans, so I had to wear mine over the pants. For a very few well-seasoned miners, this worked fine, but it wasn't a good look for a new guy. And the flannel shirt, in addition to soaking up every possible ounce of water and sweat, wouldn't last more than a couple of shifts before falling apart.

I had my new gear, though, and thought I was ready for work underground. The truth was that, other than being more than willing, I was not ready when I showed up for my first day at Kermac Section 35.

The Underground

With an armful of new gear, new clothes and boots, I reported to Shotgun's office that first day when I should have been reporting to Mel Vigil's office. That didn't go well as Shotgun, looking annoyed, just told me to go see Mel and he'd get me started. Chain of command. Should have known that.

Mel was OK with me, although he still had that kind of bemused look. He told me I was assigned to Frankie Garcia's crew and to find Frankie and let him know who I was, after which Frankie would assign me to various jobs. But first I needed one of those baskets, so Mel told me which was mine.

Leaving Mel's office, I went out into the Dry, which was really alive with activity as a hundred men were busy changing into their work gear. I was told where Frankie Garcia was, so I found him and told him who I was. He gave me a token with my name stamped on it and instructed me to put myself on the out board, get a charged battery and headlamp from the charging rack, and then get changed. After that I was to report to him in the lunchroom on the 1–5 level.

Everyone who goes underground has his name punched or written onto a small metal token that is then hung on the "in" board or the "out" board. This signifies whether you are in the mine or out of the mine and helps to keep track of where everyone is. This is done underground too by way of a board in the lunchroom of whatever level you work on.

Next I went over to the charging rack where large batteries were stored and charged. The battery attaches to the miner's belt. A cord with a headlamp attached is connected to the battery, and the headlamp is connected to the miner's helmet.

At the end of each shift, workers return their batteries to the

The Underground

Mine entrance in the Ambrosia Lake area just east of NM 509. Mt. Taylor is in the background. The view is the same one I had each day on my way to Section 35 (photograph by Marti Martienssen).

charging rack, where it takes a few hours for them to recharge. Many times batteries didn't fully charge, leaving the user literally in the dark. More on that later, but for now just know it's not a good thing to experience.

Tagged out and with battery and lamp in hand, I found my basket number. As luck would have it, my basket was located right between two of the most experienced miners at Section 35, Dobbs and Gibson. If I ever did know their first names I've long since forgotten them, but I only ever called them Dobbs and Gibson.

I was very fortunate really that Dobbs and Gibson, besides being highly skilled and experienced longtime miners, also happened to be tolerant of new guys, as they were both quite friendly that day in explaining to me how to raise and lower my basket.

It didn't take long to change out into my new, never-used gear that presented quite a sight, I'm sure, with my new clothes and boots, shiny hard hat, and miner's belt. As an extra added bonus to the amusement of the seasoned hands, I was wearing my brand-new, black, thick-rimmed safety glasses that, although very useful and adding significantly to safety, looked preposterous on a new guy and were rarely seen on anyone with experience.

It didn't take me long to realize the proper way to wear safety glasses is to fasten wire around the arms and hang them from your neck—sort of like a librarian but much dirtier.

Of course a new hand discovered none of this until he walked out for the first time to the hoist, where he, in his greenhorn outfit, becomes the center of attention, or the exact opposite of what every new hire would aspire to be among a group of experienced miners, helpers, laborers, and others.

The worst of it was that because I was the only new hand my first day, there wasn't anyone who could look more ridiculous than I did to deflect at least some of the unwanted attention. I definitely contrasted sharply with everyone else, from the most experienced hand to the guy in the middle of his second week.

Experienced hands wore hard hats that had been hit many times over on a daily basis. Their clothes were a kind of universal dark gray with the oil, grease, and grime so ingrained into the fabric that no amount of washing could get them clean.

All the experienced men wore their miner's lamps draped around their necks until they got underground, where they would attach them to their helmets. Their miner's belt usually sagged to one side partly because of the combined weight of the lamp battery and self-rescue device and partly out of a sense of a miner's fashion. Yes, it was quite a contrast to see the old and new standing around waiting for a ride down.

In retrospect I should have been fairly upset with my pal Greg Hornaday for not telling me all the things I needed to know about what an underground worker needs to look like and wear. Later in my career, I found the look of the new guy to be one of the funnier, more entertaining sights we would see above ground, so I was pleased to have provided some comic relief to my associates during those first few days. So thanks a lot, Greg, but OK, I get it.

I'm sure that when I showed up by the hoist, someone elbowed someone and said under his breath something like what I always later

The Underground

said: "Oh my God, pard, look at this guy. What the hell is this?" What the hell is right.

I was, however, humbled to have suffered the burden of the new hand alone that first day, because it seemed from then on that first-time hires seemed to show up in groups of three or four. There they were, having found one another, huddled by the hoist in their own small group, one looking more absurd than the other. I always imagined I wouldn't have been the most ridiculous looking of my group, but I'll never know.

The appearance of a new hand waiting for his first ride down into the mine was certainly entertaining, but what I think most of the old hands found most amusing was their knowing what the new hand would be going through those first few days and weeks underground. Mercifully the old hands mostly stood quietly observing, being keenly aware of the angst the new guy was experiencing. It wasn't a feeling you would wish on anyone but something the new hand had to accept and endure.

Later on when I saw new hands standing around on the surface, I always wondered if they'd make it. I do not so much mean whether the person could do the work, but whether they would panic when they got off the cage, and head right back up to the surface as sometimes happened.

I might have looked ridiculous the first day I showed up, but I was going to make it. I was sure of that.

Earlier I mentioned the headframes, and now I was getting my first close-up view of them. They sat directly over the main vertical shaft into the mine.

In the Ambrosia Lake area and in many other mining areas around the world, these are large triangular steel structures of fifty feet up to over two hundred feet. Although they can be made of wood, such as what was used in the earliest days of mining, or of concrete, at Ambrosia Lake, they were predominantly steel.

Depending on where a miner came from, I would hear the headframe referred to as a hoist frame, headstock frame, or simply the shaft. The function of the headframe is to provide support for the hoisting of people and ore. Large, heavy cable lies over wheels on top of the headframe, running back a couple of hundred feet or more to a large

Underground and Radioactive

drum located in the hoist room, over which the cable is wound or unwound. Usually a single operator, or hoistman, is in charge of operating the raising and lowering of personnel, materiel, and ore by a system of clutches and levers.

I knew some hoist operators, and they were very well trained and highly skilled. Most of the lifting and lowering was done blind by bell signals sent by mine personnel, either by the cage area or underground, to the operator in the hoist room.

The cage is akin to an elevator car in that it carries people up and down a vertical shaft but with a much more industrial, utilitarian appearance. It really did look similar to a cage, being made of metal bars and angle iron. It had a metal grate floor and a sliding door that looked something like a jail cell door. The cage held probably fifteen to twenty workers tightly packed, true to its name, when I rode in one I felt caged in.

I don't like crowds much, so I never enjoyed the cage, but it was the only way into and out of the mine.

It was a helpless feeling standing there in the cage having no control whatsoever, dependent upon the skill of the hoist operator. I always hoped that things were going well at home, that he had slept well and was in a good mood. Needless to say, that wasn't always the case.

There were many times during a ride down into the mine that the hoist operator would drop us twenty or thirty feet before recovering. Always a nice surprise. Those drops were quite the ride and elicited some colorful discussion among those on the cage concerning the hoist operator. I suspect the drops were sometimes not accidental but couldn't prove it. More than once drops were so severe that individual miners made it a point to get in touch with the hoist operator after a shift change, and a not-so-pleasant exchange ensued. Fortunately, the ride down was usually a nice, smooth descent.

The drops and other cage experiences were all in the future. For now, I was just standing out there, all shiny and new, with the other men waiting for a ride down to the 1–5 level.

A lot was going through my mind then, but I never doubted that when I stepped off that cage for the first time, I'd be accepted into the subterranean fraternity of miners, motormen, welders, engineers, geologists, and others, or in my case laborers.

The Underground

Headframe of the type common to most mines in the Ambrosia Lake area. Large wheels are visible on the second and third deck over which cable runs back to a hoist room where it is spooled on large motorized drums. One set is for raising and lowering people and supplies and the other is for lifting ore out of the mine (photograph by R.D. Saunders).

I was looking forward to learning all about mining and working with a lot of people who were earning a tough and honest living. I was much more concerned about being able to do the actual heavy physical labor than I was about getting along with my coworkers or being confined in the relatively small working areas of a mine.

I think I deferred to just about everyone before I stepped onto the cage for the first time until those left waiting were myself and a few relatively new guys.

We did finally step into the cage for the ride down to the 1–5 level at about eight hundred feet. One of the men knew the bell signals, and off we went. It was a great ride, as I remember, as I was excited and full of anticipation, and thankfully we weren't dropped.

I came to know a number of people who stepped off the cage, took a look around, did an about-face, got back on, and were never seen underground again. Most often they ended up working at a support

Call bell at the station used to signal the hoist operator on the surface. The sequence of the bell pulls indicated to the hoist operator whether those on the cage wanted to go up or down and to what level of the mine they were headed. I used this system so frequently that I can still use it from memory (photograph by R.D. Saunders from an exhibit courtesy New Mexico Mining Museum).

job on the surface, driving a truck or loader, stocking supplies, or something similar. And quite the contrary to what I'd heard in orientation, some miners had at least a degree of contempt for these individuals. I wasn't one of them. I knew there were a lot of jobs I wouldn't or couldn't do myself, and whatever the equivalent to stepping off a cage is in those jobs, I would have gotten back on and gone back up myself.

As time went on, I came to feel sorry for the surface guys—that they weren't able to savor the grand ecstasy of the absolute blackness of the underground, the muffled (or sometimes not) explosions, and the unique smells. It was their loss that they never got the opportunity to expand their vocabulary through working relationships with miners

The Underground

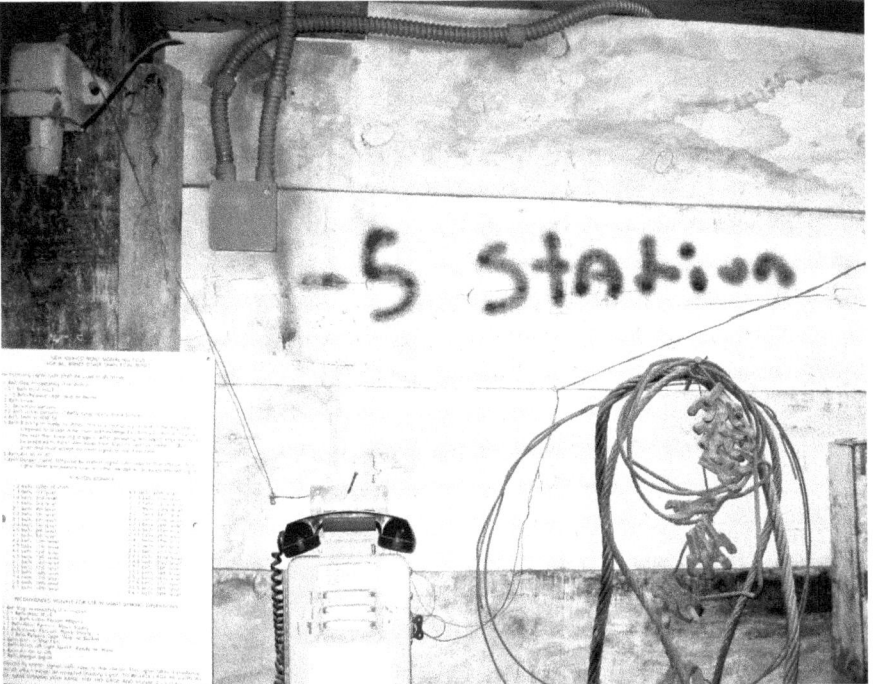

The 1–5 level station. There was nothing fancy about life underground, as the signage illustrates (photograph by R.D. Saunders from an exhibit courtesy of the New Mexico Mining Museum).

from all over the world willing and able to teach obscenities in a dozen different languages.

The cage came to a stop on the 1–5 level at an area called the station. The station is one of the few very well-lit areas underground and is large by mine standards, being maybe fifty by fifty feet with a ceiling of twelve or more feet high.

Stepping off the cage, I noticed right away that the walls were lined with wire-mesh fencing material that was bolted on. I soon learned that walls are called "ribs" and the ceiling the "back." The bolts are simply called bolts, or rock bolts specifically, and work on the same principal as plastic expanders used with screws to fasten objects to Sheetrock walls.

All kinds of supplies were stacked in the station: large stacks of

Underground and Radioactive

timber, rock bolts, wire mesh, and everything else used underground. I would soon learn the specifics of those supplies I now saw all piled up there in the station.

I played follow-the-leader as our group headed to the lunchroom. The lunchroom is another fairly large cutout into the rib, large enough for a twelve-foot picnic-type table and a small lectern used for announcements. It was more or less the bosses' station and office. There was a sign-in and sign-out board, a blasting board, and various rescue equipment leaning against the walls.

The walls were covered with maps and diagrams of the 1–5 level. I'd say the roof was about eight feet or so. It was quite a comfortable place, really, and one where everyone gathered before shift, for lunch, and after shift. When the bosses weren't wandering around checking on their crews, it's where they spent a lot of time.

I had no trouble finding Frankie who was prominently seated at the bosses' station. Frankie was another well-experienced former miner converted to shift boss either by choice or necessity. By his countenance he wore every bit of his experience. To me this guy looked just like what I imagined a miner might look like. I have to say I felt silly standing there before him in my shiny new outfit.

Unfortunately, it quickly became apparent that Frankie had been through my personnel file and knew where I'd been and how I'd come to show up there in front of him. It didn't appear he'd been much impressed with what he'd seen.

He probably said hello, but I don't remember hearing it. What I do recall is he immediately began his personal Section 35, 1–5 level orientation session.

Opposite, top: **Underground lunchroom. In addition to the lunch hour, workers gathered here before and after their shifts when bosses would relay announcements and reminders that were usually safety related. It was one of the more comfortable and safe areas of the mine. In the right hand corner are a rescue board and basket used to transport injured workers.** *Opposite, bottom:* **Bosses' station in the lunchroom. If they weren't wandering around the mine, bosses could usually be found here. The lens cleaning station was where mine workers could clean their safety glasses. Because they fogged up easily and otherwise got in the way, very few workers wore them on a regular basis (photograph by R.D. Saunders from an exhibit courtesy New Mexico Mining Museum).**

The Underground

Underground and Radioactive

"Now, I don't take no bullshit. You get here on time and ready to go every day, and you work with who I tell you. You don't know how to do anything, but we'll fix that. You'll be getting the shit jobs, and don't complain because I don't give a fuck."

He went on to further explain that light was important underground. "And don't get no damn dead battery 'cause I don't want to be looking for your dumb ass."

That ended my personal orientation session. Not real detailed or specific, but just enough to get the gist of his opinions. Now on to whatever job he had in mind that I assumed couldn't be good.

As a new guy, I was initially referred to as a greenhorn by many miners and laborers. No problem there, as I knew what a greenhorn was from the many Westerns I'd watched growing up. It was not much of an advantage, but at least I knew what it was.

Real miners habitually thought of us laborers as people from which to keep their distance. It wasn't that they disliked them personally, but avoiding inexperienced hands reduced the likelihood of injury as a result of some ill-advised act by the laborer.

It was a wise policy, as I found myself making a whole bunch of mistakes early on that could get someone else hurt. Even so I found during my first few weeks underground that miners did occasionally generously acknowledge my existence with some kind, inoffensive greeting like "Hey, pard." Beyond that they let it be known I would have to earn anything further through hard work and experience.

As a greenhorn laborer, I'd be doing anything for anybody at anytime, anywhere, and under any conditions and sometimes the worst conditions—jobs nobody else wanted, that were dangerous and didn't pay all that much. And by the way (I was told), don't be asking any questions about it.

If I turned out to be a decent worker, I'd move up and begin earning respect. Good laborers had a reputation for staying out of the way, keeping quiet and never costing anyone anything or hurting anyone through their inexperience and ignorance.

Most new laborers spent a lot of time loading and unloading supplies from the cage that, after dropping off a shift of workers, immediately became a cargo carrier. Laborers would stack the materiel in the station area and then load whatever it was that was needed back in the mine onto a motor (more or less a small locomotive) that came to the station. So it was that Frankie immediately assigned me to start unloading supplies as they were delivered from the surface.

It was, as I'd anticipated, very heavy work.

This mine used a lot of timber, which I'll get into later. The worst of it was the very unwieldy and incredibly heavy twelve-by-twelve inch twelve foot posts that we needed two men to lift and stack. My being inexperienced at stacking huge timber resulted in several collapsing piles. More than once I was thankful for the steel-toed boots.

There were stacks of rock bolts that I later found to be kind of enjoyable to install but not much fun to unload and stack by the thousands. In short everything that was used in a mine came down by way of the main shaft, and it was the laborers who saw it all first.

Rarely in our mine did anyone ever work alone. That was a safety precaution, and it worked. On occasion one partner would sustain an injury and the other would go for help. Without the partner the injured worker could lie there for hours before either a shift boss came by or he was noticed missing at lunch or at the end of the day. It was very rare that I ever did much alone during my time at Section 35.

Frankie paired me up with a young guy by the name of Grimm. I can't recall his first name, but he was relatively new, having been at the mine for just a few weeks. Learning that Grimm had been working at the station that long was a little depressing, as I thought that would be my future too. But as it turned out, there was something else keeping Grimm at the station, and whatever it was I never found out about. It couldn't have been his work, because from what I saw he worked hard and did everything that needed to be done.

Frankie was seldom to be seen that first day, and in fact, after he had given me his speech and told me what to do, he had vanished. I think he was much more interested in visiting the miners he was supervising than a couple of greenhorns stacking lumber.

About four in the afternoon, people from all over the 1–5 level came straggling back to the station, reporting whatever progress they had made that day to the shift boss, then riding the cage up to the surface. Once again Grimm and I, as the newest men, didn't want to be first into the lunchroom or first on the cage, so we took our time and, when it seemed reasonable, went into the lunchroom to check out.

Underground and Radioactive

Still no sign of Frankie, so I thought we should wait for him, but Grimm thought we should head to the surface. If Frankie wanted us, he knew where we were. So, moving our tags from the out board to the in board, we headed for the cage and the surface. I was dog tired, my shirt was falling apart, and my hands were raw, but I'd made it through the day.

Those first few days were exhausting. I had the wrong clothes, of course. My flannel shirt, in addition to becoming soaked with sweat, couldn't take the work and ripped right away, first at the elbows and then just about everywhere else. Cool air rushes down the main shaft to the station, and with my soaked shirt on, I was more or less cold all day long.

Everything was just so heavy, and it didn't help that I had forgotten my gloves. My hands immediately began blistering, and aching constantly made for an initially trying existence.

By working at the station, I soon learned the unique mining names given to some of the more common items that we were unloading. Jackhammers were chippers, shovels were muck sticks, rock drills were machines, and sledgehammers were double jacks. I don't know why, but a pick ax stayed a pick ax. With rock bolts it depended on what they were to be used for. A bolt used to help stabilize rock overhead in the back was called a roof bolt, but it could also be a rib bolt or a block bolt.

A jack similar to an automobile jack but considerably larger and used by miners as a temporary means of ground support was called a roof jack. Those two instances were the only times I heard the back referred to by the word roof underground. Why they weren't called back bolts and back jacks I never knew.

Sometimes a laborer's first job at the station would go on for days or weeks, depending on the needs of the boss, the laborer's ability to work, and the boss's opinion of him. I knew if I did well, I would eventually be allowed to accompany the motors back to the working areas of the mine. I would then unload the supplies and accompany the motor back to the station to load more supplies, riding comfortably in an empty ore car all the way. As it turned out, my working exclusively in the station area lasted only for about a week before I was needed elsewhere.

At the beginning of my second week, Frankie let me know I'd have

The Underground

to ride the motors along with the supplies I'd loaded back to the work areas, where I would then unload and deliver them to the miners. In my mind and from what I'd seen up until then, an assignment like this meant I was making progress, because now I'd come into contact with miners' helpers and every now and again an actual miner. It was lucky for me because after a long first week, I was already sick of being stuck at the station all day long.

Underground trains, cars and locomotive combined, were simply referred to as motors and the drivers called motormen. They looked like something between an amusement park train and a miniature freight train. They were fairly low profile and ran on narrow-gauge track. Attached to the locomotive were any number of ore cars capable of holding at least a ton of ore each.

Because there was no place for the train to turn around, the cars always went into the mine first, followed by the motor. What made running one of these trains difficult was that half the time the motorman couldn't see where he was going. The poor visibility made it extremely difficult for the motorman to see people or debris on the track ahead of the cars.

Debris was a constant hazard, mostly due to falling rock, but sometimes from supplies left on the track by inexperienced laborers. As a result, derailed cars were a problem but something a good motorman could quickly remedy.

You might think that a motor with ten or more ore cars attached would make a lot of noise rolling down the track and would be easily heard. That wasn't the case, though. With all the other ambient noise in a mine, a rolling motor was just peripheral sound that could easily be missed. Lights were therefore used in safety systems designed to keep workers from being run over by motors.

As a motor travels in reverse out of the station and into the mine through tunnels, called drifts, the motorman would come upon a series of lanyards hanging from the back at various intervals. Each time the motor passed under a lanyard, he was supposed pull it, setting off a series of lights flashing along the drift ahead warning anyone on the track that a motor was coming and to move quickly out of the way. Not a bad idea, but the flaws in that system were that bulbs did burn out and did not get replaced, and motormen in a hurry sometimes failed to pull the warning light lanyards.

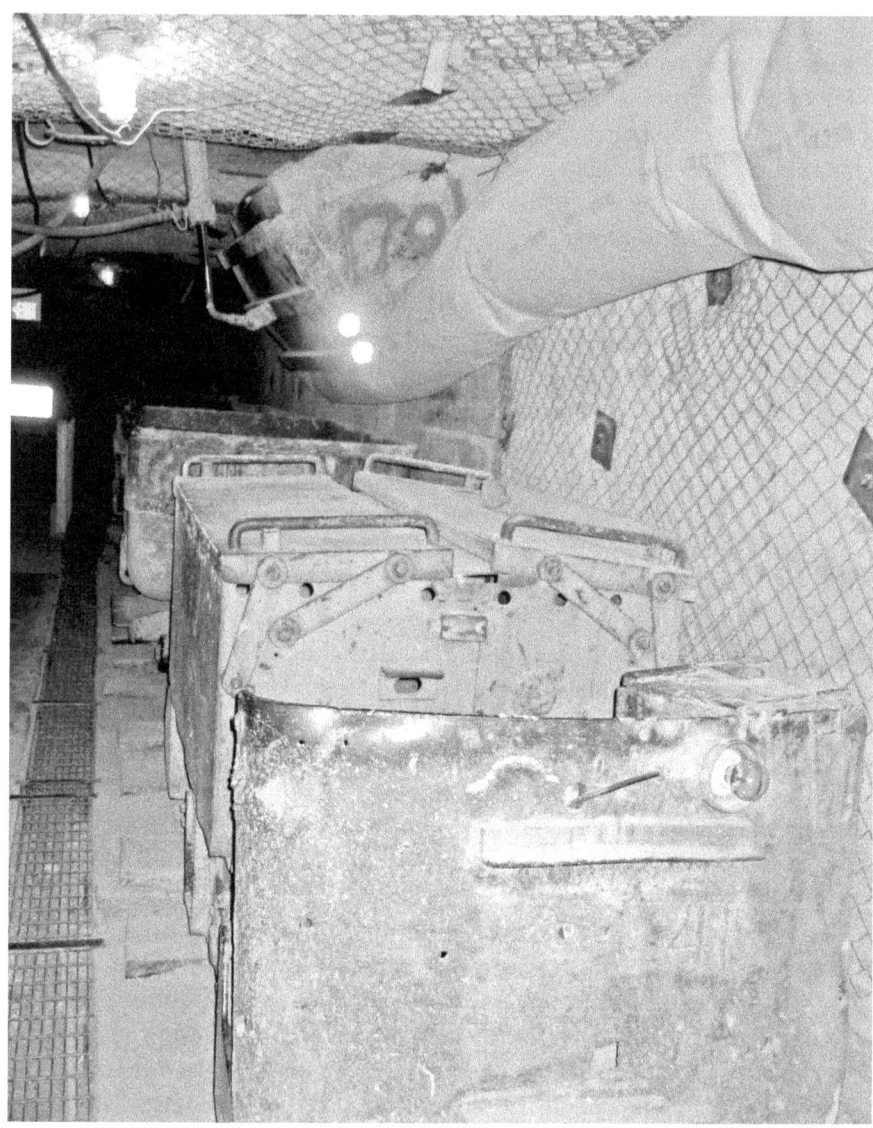

Able to haul many tons of ore, these relatively small yet powerful machines were all electric so as to eliminate the smoke otherwise generated by an internal combustion engine. The battery covers are visible just behind the small operator's cab (photograph by R.D. Saunders from an exhibit courtesy New Mexico Mining Museum).

The Underground

Because drifts were commonly too narrow to accommodate the combined width of a worker and a motor, cutouts were put in at intervals along the track drifts. When workers walking down the track saw the lights above start flashing, they would quickly make their way to one of them. This system of lights also kept trains from colliding. As with many other safety systems underground, it was simple and effective. And, as with many of the safety systems underground, it was just as often ignored.

Experienced motormen were good about pulling those lanyards but not perfect. They sometimes overlooked the lanyards or, because they were speeding, missed them altogether. In anticipation of that eventuality, there was a second fallback system, the trip light.

Attached to the back of the last car on the motor was a small flashing light. This was not very bright but didn't have to be because any light in total darkness is a lot of light. If they were working and the person on the track was paying attention at all, a trip light was hard to miss. The principal flaw of the trip light system was a dead battery and therefore no flashing light.

If a worker walking down the track was being alert, he would see the trip light in time to run to a cutout. Sometimes the worker would see it at the last moment, leaving just enough time to hop up on the coupling of the approaching car and ride it to safely when the motor came to a stop. What usually ensued afterward was a heated discussion between the motorman and the worker about safety. I rode many couplings during my time underground.

Occasionally a worker walking down a drift would not see the trip light, either because he wasn't looking or because the light wasn't working. In that case the worker was either hit or run over. Considering the safety systems used in running a motor, workers were hit a lot more than I would have expected.

I'm certain the powers that be weren't too pleased when someone was hit by a motor, but the attitude among us underground was that it was too bad, but accidents do happen. Yes, the trip lights should have been working, and sure, the lanyards should have been pulled, and ideally a burned-out bulb should have been replaced. It's a shame those things happened, but they did.

An additional flaw in the motor system was that sometimes people who shouldn't have been operating them were operating

Trip light. Although in this example it is not, the light was attached to the rear of the last trip of ore cars proceeding into the mine. This safety device saved me from being run over a few times. Seeing the flashing light coming at me, I either ran the other way or jumped onto the coupling and went for a ride (photograph by R.D. Saunders from an exhibit courtesy New Mexico Mining Museum).

them. I was one, but this also included many miners and miners' helpers.

Because both miners and motormen got paid for cars of ore hauled, there was a delicate balancing act between motormen keeping miners happy by hauling ore and miners keeping their ore chutes full for the motormen. Often a miner would have a full chute of ore but no motorman around to empty the chute into ore cars. Something like that brings production to a halt for the affected miner, who is getting paid only for production, not by the hour. It was common practice for a miner or his helper to go get an idle motor and do the ore hauling himself. Basically what you had were a bunch of unqualified drivers on the track, which led to a lot of unpulled lanyards, dead trip lights, spilled ore, derailed cars, and more.

Motormen having their salary determined by the number of cars of ore hauled were not at all happy at the prospect of being assigned to hauling supplies, which paid them relatively little.

When it happened that a motorman had been assigned to haul supplies, he would be in a bad mood and in a big hurry to get the supply run finished. So while I was pretty happy about being able to accompany the supplies back into the mine, I usually did so with a grumpy motorman who wasn't so happy about waiting around for some slow-moving, know-nothing greenhorn.

So began a regular routine for me of accompanying supplies back to working areas of the mine. I loved going back into those areas and getting a feel for it, and I can't think of a single thing I didn't like about that right from the start.

As I said, the motormen didn't have any particular affection for either hauling supplies or for laborers in general and were not shy about withholding their displeasure. They were constantly on me to get the supplies on and off the motor as quickly as possible so they could go back to hauling the ore that made them money.

Upon reaching a delivery destination, what usually happened was that I would dump supplies along the track as fast as I could and as close to the manway and raise as I could.

The manway was a vertical passageway with a ladder, sometimes a very long ladder, leading to a production area called a *stope*. Usually, but not always, there was another vertical shaft called a raise next to the manway used only to lift supplies into the stope.

When I came to know some experienced miners, I started asking why the production areas were called stopes. Nobody knew, and I got a lot of unwanted attention for even asking such a thing, so I started looking around for any information I could find. My own feeling is that it is unclear exactly where the term originated, but it is likely to have come from hard-rock metal mining.

High-grade metals like gold and silver often are found in uneven, diagonal veins. As the miners follow the ore up, they do so in a series of steps. Gradually the area being mined assumes the shape of a flight of stairs. *Stope* then became a derivative of *step*.

Underground and Radioactive

On the other hand, I did a lot of stooping during my time underground, so for me the term *stope* could just as easily have been a variant of *stoop*.

All those supplies couldn't stay on the track for long, so it was my job to assist in getting the materiel up to the stope. More often than not, I would work with the miner's helper to accomplish this task. We did it with a winch installed at the top of the raise. The helper would send the cable end down with a clevis attached, and I would fasten and cinch up everything as best I could. Of course the helpers weren't in a great mood either, having to stop production to haul up supplies, so they wanted it done as rapidly as possible.

With no experience attaching anything to a clevis, I would sometimes find supplies coming crashing down the raise. I soon learned to never look up a raise as supplies were being lifted. If it happened often enough, meaning more than once, I'd find myself face to face with the miner's helper, being asked in a colorful way what kind of clown Frankie had sent to deliver supplies.

Then there was Frankie himself to occasionally contend with. He did manage to materialize on a few supply runs, and he too sometimes was wondering what clown he had hauling supplies. In this way I soon learned to haul and deliver supplies quite efficiently.

Soon, and I put soon at about a week, I was doing a very efficient job delivering supplies to the work areas. The people who mattered seemed to notice and appreciate it, in that I wasn't being cussed at nearly as much. It was around this time that Frankie began calling me "pard," which I interpreted as more personal progress.

One motorman suggested to Frankie that I drive my own motor full of supplies, saving him, the motorman, valuable time and thus increasing overall mine production. Frankie thought this was a great idea, so I was told to use an idle motor to drive the materiel back myself. Of course my having no idea how to run a motor made little difference to Frankie. I was just expected to do it.

If I could remember who it was, I would have someone to blame by name for what followed, but I don't.

There was an idle motor with a few cars sitting just beyond the dump station, where ore is delivered to hoppers that take it to the surface. It turned out to be quite an easy machine to operate. It had one

The Underground

Standard ore car. Ore was easily dumped into the main collection chute by flipping the car via the four hinges visible just above the wheels (photograph by R.D. Saunders from an exhibit courtesy New Mexico Mining Museum).

forward and one reverse gear and was all electric. How much pressure was applied to the handle determined the speed of the train.

Couplings on ore cars are the same as on any train—not real difficult to figure out, but it did take a certain amount of finesse to make the connections by touching at just the right speed. That was the real learning curve: hit a car too hard attempting to couple, and it could and usually did make the car hop off the track. I managed to derail many ore cars trying to figure out the finesse component.

Ore cars were not light, so when they derailed they couldn't simply be lifted back onto the track by hand. They could be jacked up and, when high enough to clear the rail, simply nudged back onto the track. At first that's what I did, but it was time consuming and frustrating.

I'd noticed another method used by motormen that was much easier and considerably faster. It involved placing a wedge under the derailed car at a slight angle to the ground. The motorman would then use the motor to gently push the car, which would then lift itself onto the wedge, after which the motorman would run back and just push the car back onto the track. It was a simple system that any good motorman could use. Through my many derailments, I soon learned how to do it.

After managing to get my ore cars attached to one another and the motor, I moved out of the dump station. That's where I learned about switches.

Exactly as on any railroad, the mine track had a series of switches to move ore trains from one track to another, depending on what drift a motorman needed to go. There were also switches near the station and the ore dump depot. They were fairly easy to operate, but without frequent maintenance they didn't work properly, and then an ore car or a motor itself would derail.

Through a series of derailments, I did learn how to properly maintain and use the switch system to get my train back to the station where supplies waited to be delivered. I was very fortunate never to have derailed a motor or a full ore car, as due to their weight they were extremely difficult to get back on track.

While I was learning these valuable lessons, I had miners' helpers running back to the station to find out what was holding up delivery. Soon I was delivering supplies back to the stopes with fewer and fewer problems.

Just as I got really good at supply delivery, there were new hands showing up to work in the station area, so Frankie decided I should be back in the mine performing other tasks. First new task: latrine duty.

Mine sanitation was interesting in that it was more or less minimal. If a worker needed to relieve himself, he stopped and did so anywhere. No problem. On the other hand, if the relief included other than urination, there were latrines built for that purpose in cutouts along the main track drift. Lovely areas.

When a cutout latrine was no longer serviceable, it was simply filled in, and a new cutout was constructed further down the drift. So my new job along with Grimm was to fill in old latrines. There was a lot of dirt, sand, and rock around, so it was easy to accomplish.

The best part about this job was there were a lot of latrines in

The Underground

Ventilation tube with a blower attached. These devices brought fresh air into any area of the mine, lessening exposure to radon gas. I always made certain I worked with one of these units and the attached vent tubbing close by (photograph by R.D. Saunders from an exhibit courtesy New Mexico Mining Museum).

Section 35, so I started to learn my way around. I began to appreciate trip lights too, as several times during this assignment I was either running for a cutout or riding on a coupling. Fortunately, latrine duty was short lived, and I moved on to more exciting work.

Frankie sometimes assigned Grimm and me to hang vent tubing along the track drifts, up manways, and into stopes. The tubing was made of a light, flexible, reinforced yellow plastic material and was about three feet in diameter. It was attached to a blower used to get fresh air from the main track drift back to otherwise unventilated areas of the mine.

Vent tube was required in all production areas of the mine, and nobody was supposed to work further than several feet from one. "Take your vent tube with you" was something every boss told miners as production advanced in their stopes.

Hanging vent tube slowed down miners and didn't pay much, so they were not especially good at following that rule. If Frankie, in making his rounds, noticed insufficient vent tubing, then Grimm and I would be dispatched to the area to correct the problem.

I had to experience an unventilated area only once to learn the value of vent tubing. It was an uncomfortable, stuffy environment that I refused to work in if I could help it. When I had my own production areas, I never worked without a vent tube close by.

For me, vent tube hanging didn't last long, though, as more exciting adventures were on the horizon.

Moving Up

I had been underground about four weeks by now. I had learned a lot, mostly through the colorful encouragement of irate bosses and miners and by trial and error as opposed to actual instruction, of which to date there had been approximately none. That was just the way things were done with new hands. Either learn the job, show you can do the job, or get out. Some got out. I hung around.

By this time, I had gotten some new, durable mining clothes including, denim overalls and a denim Wrangler shirt. The overalls had a lot useful pockets and were tough, as was the shirt. My clothes had already taken a beating, so much so that I felt I was beginning to look like a miner. One thing about well-worn, dirty, and beat-up clothing was that it was incredibly comfortable to wear. My helmet was getting nice and beat up too and was no longer shiny. All in all, I was starting to look the part of an underground worker.

I was feeling more confident too. My strength and abilities had improved considerably, so much so that I sometimes felt a little cocky about what I could do.

Frankie pulled me and my partner aside one morning just outside the lunchroom and explained that one of the pneumatic doors on an ore chute was stuck.

"The door on 907 is stuck closed, and I want you two guys to get it unstuck," he said. "I don't care how you do it, just get the chute open."

He went on to explain that he thought the chute was full of ore, but he wasn't sure, and he added a harbinger: "If the goddamn chute is full of water, it'll flood the drift. If it is, don't let it all out at once. If it's full of ore, get back to the station and get a motor to pull it." Then looking directly at me, he said, "You're in charge, pard."

I was on my second partner now. Grimm was still back at the station, and I was now working with Anthony Gonzalez. Anthony was around eighteen or nineteen, and I'd seen him, knew he was more of a greenhorn than I was, but didn't know him at all.

There was a rumor floating around that he had been arrested going 150 miles per hour on I-40 in his new Pontiac Firebird, so the first thing I did was ask about that. Sure enough it was true, and he no longer had a license. Although I was not in the market for a new Firebird, I knew where I could get a good deal on one.

Frankie, saving his best surprise for last, walked us over to a powder box by the cage door. Handing me about ten sticks of TNT and several fuses, he said, "Use this if you have to."

I accepted the TNT and advice with the most nonchalant posture I could conjure up and a nod of my head, all the while thinking, I'm in charge? I was amazed that Frankie thought I knew anything about TNT.

To the first question: sure, being in charge made sense. Anthony, as a recent high school graduate and only a couple of months out of school, had started at Section 35 a couple of weeks after I had, making me the senior laborer.

As to my lack of knowledge of TNT, I think Frankie was well aware of it but didn't care if I knew anything about it or not. He just expected the job to get done.

At the time I didn't know it, but the stick form of TNT, universally referred to underground simply as *powder*, is an extremely stable explosive made up of a silica-and-clay compound soaked with nitroglycerine. It takes a small but powerful primary explosion, in the form of a blasting cap, to set it off but can otherwise be thrown around or used to pound nails, for that matter, without fear. When it was stored too long in a powder box, it could liquefy, and that, as I found later, presented an unpleasant predicament.

When Frankie Garcia handed me that roll of fuses and box of powder, the wise thing to do would have been to emphasize to him the obvious: I had little knowledge of what these things were or how to use them.

Regrettably I stayed silent but was thinking, Frankie? I, uh, really don't know anything about, you know, dynamite. Not emphasizing that fact more emphatically wasn't a fatal mistake but could easily have been.

Frankie's motivations aside and despite my own misgivings, I was proud to be in charge of something this early in my underground career and was determined to get the job done. Anthony and I would get that ore chute open no matter what.

I had hoped that Frankie, at a minimum, would have taken us back to the chute in question, but he turned away and walked back into the lunchroom without further comment. With that, Anthony and I headed down the main track drift on our way back to the 907 chute that might be full of water or might be full of ore.

As we were making our way down the main drift, I kept thinking of Frankie. What in God's name is wrong with this guy? Does he really not know how clueless we are? Is he hoping we blow ourselves up, or is this how you learn to be a miner?

It was true that I'd gone underground in part to be able to legally blow something up, so I was excited, but this was ridiculous. Working around powder and learning through trial and error was really not the way to go, because we were probably only going to get to make one mistake with it. Luckily enough that turned out to not exactly be true.

All along the main track drifts were chute openings, over which was an eight-to ten-foot diameter tube bored vertically up to a stope. These could be short tubes of ten feet on up to two hundred feet.

The ore chute was a hydraulically operated apparatus. A motorman pulled an ore car up to and just under a chute, then pulled a lever that opens the chute lip and released the stacked-up ore. When the car was full, he closed the chute lip, pulled the motor ahead, and repeated the process.

Occasionally a chute would jam and refuse to open. Usually this was due to rock being wedged in at just the right angle to stop the door from opening. Once the chute was full of ore and the door was jammed closed, production would stop until it was fixed. That was our assignment.

Every stope was given a number: the farther back into the mine, the higher the number of the stope. Even for us, finding 907 was easy. All the chutes that ran the length of the main track drift had large white identifying numbers painted on the sides, making it easy for the motormen to identify them.

When we arrived at our stuck chute, I immediately noticed

Underground and Radioactive

dripping water. There was a slight opening, and I could see there was also some ore in there.

Apparently it had been some time since 907 had been in production, so nobody knew if the chute was full of ore or full of water. In retrospect I think someone should have known before sending a couple of inexperienced knuckleheads in to find out.

It wouldn't have been that difficult to shine a light into the chute from above or just drop a rock down and listen for a splash. Either nobody thought of that or they were too lazy. I should have known to do it myself, but I was just focused on what Frankie had assigned me to do.

I stood there in the drift, assessing the situation. If the chute really was full of water, it would be a lot of water, considering it was a ten-foot-diameter tube running 150 feet high.

On the other hand, maybe the chute was mostly filled with ore. Had I known then what I later learned about miners and ore production, there was close to zero possibility that any miner would have left a 150-foot chute full of ore without getting paid for it. I would have then assumed it was full of water. But at the moment, I knew none of that.

My instructions were to get the chute to open. Frankie had made it clear that should the chute be full of water, under no circumstances were we to let the water flood out, as it would swamp the drift. If there was ore in there, a little on the track would have caused no harm and could be cleaned up quickly.

It seemed that Frankie forgot that in addition to flooding the track, should all that water flood out it could quite possibly drown us. But at the time, Anthony and I hadn't thought about that. Our instructions seemed clear enough, and as new laborers we didn't ask any questions of our shift boss.

Trying the chute lever confirmed that the chute lip was stuck, as the door never budged. Along with the powder, all we had were a couple of pipe wrenches and a double jack. It was obvious that a puny pipe wrench would be of little value on this job, so we figured if we pounded on the chute with the double jack, it might loosen up enough to where we could work on prying the door open. Although we had nothing to pry with but our pipe wrenches, this seemed logical. So we took turns pounding on the chute and kept trying the lever, but the door never moved. It was then that I turned my attention to the powder as the solution.

My previous experience with powder and its uses consisted of watching television western reruns, from which I gathered you stuck a fuse into the stick of dynamite, lit it, sometimes threw it, and always ran for cover, diving if necessary.

That was a big decision that I had to think on, so I pulled out my pouch of Red Man tobacco for a chew. A lot of guys used chewing tobacco underground. Smoking was prohibited by federal regulation, and while many miners ignored that rule too, most did not, so if they smoked, they needed a substitute. A real advantage to chewing underground was being able to spit anywhere. After all, how much dirtier could it get?

Many miners who chewed used a plug—that is, tobacco that is compressed into a small, hard square. To use it you'd bite off a piece and put the remainder back in your pocket. Loose-leaf tobacco came in a foil pouch from which you'd take a smallish pinch for chewing. I preferred the leaf tobacco in the foil pouch and always had some handy and stored in the top front pocket of my overalls.

You hear a lot about the long-term implications of tobacco use, but there are short-term affects too that I can attest to firsthand. Inadvertently swallowing a chew of tobacco will ruin your day. It happened to me several times and in doing so ruined several days by causing a case of severe indigestion. Definitely do not swallow tobacco.

As I stood there chewing away, contemplating the use of powder, I wondered how far we would have to run or even if we had to run at all, seeing as I had no idea how fast the three feet of fuse lead would burn. But running out of options, I decided to try using the powder to blow the chute open.

"You ever used this before?" I said to Anthony.

"Nope, never," he said.

"Me either, but let's give it a try."

I picked up a fuse in one hand and a stick of powder in the other and said, "I think what you do is stick the blasting cap end of the fuse into the powder. We can jam the powder up between the lip of the chute here, light it, and run," I said.

Underground and Radioactive

Anthony, not seeming very enthusiastic, said, "Will we have time to get away?"

That was a reasonable question, so I thought about it for a minute before saying, "I don't know, but we'll probably have time. Let's just try it. It won't blow the door off, but it might loosen it up."

Of course I had no idea what one stick of powder would do or whether it would blow the door off, mangle it beyond further use, or anything else, for that matter, but Frankie must know, and he's the one who gave us the powder.

I balanced a stick of powder on top of the door where the moveable lip met the stationary section of the structure then stuck the shiny metal blasting-cap end of a fuse into the end of the stick of powder.

I sent Anthony down the track drift quite a little way to act as a lookout and assuage his fear of being able to get away by already being away. Using my handy disposable cigarette lighter, I lit the fuse. I was prepared to sprint away but it seemed to be burning slowly, so with no need to panic, I made my way down the drift toward Anthony a hundred yards or so away. The fuse was a lot slower burning than either of us had figured, so we ended up waiting a good five minutes before the powder went off.

The explosion wasn't as loud as I thought it might be, and while there was some smoke, it wasn't much. I was feeling proud of myself for finally having blasted something as I said to Anthony, "Let's check it out, pard." As full of confidence as I was, we nonetheless cautiously walked back toward the chute.

Not only did the door seem intact, but I couldn't see a mark of any kind where the powder had gone off. Still, it was possible that the door had been loosened. Carefully trying the door handle to see if it would budge the chute door proved futile. Still, we had successfully blasted something, and although that had accomplished nothing, we hadn't hurt ourselves or—it appeared—damaged anything. So far we were doing a fine job. Next logical move? Make a bigger bomb.

I next tied three sticks of powder together and jammed them up against the chute door and inserted another length of fuse with a blasting cap. After again sending Anthony back down the drift, I lit the fuse and walked down to join him.

Several minutes later came a much bigger explosion, this time with considerably more smoke. When it cleared we again walked back to observe our handiwork. As before there was no damage, and in trying the door lever, we found no loosening of the chute door.

I decided we were not using enough force on the door, so we'd better double the effort this time by tying six sticks of powder together.

That looked like a good-sized bomb and just might do the trick. No way the door could resist that kind of force, right? Again, I lit the fuse and again walked down the drift where I waited with Anthony.

This, our third try, resulted in a massive explosion and a lot more smoke, which took some time to clear. That, I thought, was a real bomb. No way this couldn't have worked.

As before, there was no visible damage done to the door, but it now opened ever so slightly when I pulled the handle, enough so that I could clearly see some of the ore at the base of the chute.

The chute was dripping water, but to me there was nothing to indicate that there was anything but rock filling it. Of course the prudent thing to do would have been to climb the 150-foot manway and shine a lamp down the chute from above. But that was a long climb, and having seen the rock at the base, it was clear there was no water, and besides, the 1–5 level was above the water table. How could there be much water in the chute anyway?

Thinking Anthony and I might work together on the problem, I stood on one side of the chute lip in hopes of getting a better look while he would operate the handle on the other. "Just try moving the handle a little bit," I said. "Just move it very slowly. We don't want to dump all that rock on the track." So while I stood just to the side of the chute lid looking for any movement, Anthony gave the handle a pull that was a lot more than he'd intended. That's all it took.

The bomb had worked, and the weight of all that water above forced open the chute door the rest of the way. There was no stopping it. As the water came rushing out, I was immediately submerged and washed down the drift fifty feet or so before I was able to grab hold of the wire mesh bolted to the ribs, the whole time thinking, I'm going to drown in a mine?

I had known there was the possibility of being caved in on or blown up, but drowning in a mine? I hadn't considered that.

Still holding on to the wire mesh as the torrent was rushing by, I was able to pull my head above water. Looking around for Anthony, I didn't see him but knew he was somewhere down the drift and still submerged.

The surge of water seemed to just keep coming as I clung to the wire mesh, still able to keep my head above water. Somehow in this, the first of several close calls during my short mining career, I began to think I might survive.

Finally, after what seemed an eternity but had probably been no more than thirty seconds, the chute emptied, and the water in the drift began to subside.

Down the drift no more than fifty feet, I could see Anthony had a firm grip on the mesh, drenched from head to toe but otherwise alive and well.

It's funny how quickly thoughts can turn from the most serious to the most trivial. Once my survival seemed assured and I knew Anthony was fine, I became concerned that my Red Man chewing tobacco would be ruined, and how was I going to get through the rest of my shift with no chew? Perhaps the aluminum-lined bag would save it? I checked the top pocket of my overalls and found my Red Man was ruined by the flood waters. I was despondent but not for long.

Then I noticed that my miner's headlamp, having been submerged, was still working just fine. That's pretty cool, I thought, great engineering.

I then started congratulating myself on a job well done. The chute lip was no longer stuck, the chute itself was empty, and nobody had gotten hurt. It was all working out. Not exactly.

From down the drift I could see a light rapidly approaching and then the figure of Frankie, emerging from the darkness, running toward us. I was thinking he would be happy to know Anthony and I were OK and that I could show him we had indeed accomplished the opening of the ore chute. We might have messed up and let the water out, but nothing seemed damaged, and we had lived, so Frankie would be pleased. Frankie was not pleased.

Red faced and enraged, Frankie let loose with a string of expletives unlike any I had heard underground up until then. "You stupid cocksucking, motherfucking idiots. You flooded the drift all the way down to the station. It's a fucking mess, you assholes. Get up to the fucking surface."

Obviously the chute being again operable was of no consequence to Frankie. Too stunned from the event anyway, I didn't mention it, electing instead to follow instructions and head back to the station with Anthony.

As we were walking through the drift, I could see what a mess it was. The force of the water had taken a lot of loose rock from the ribs and deposited it on the track. Nothing big, but there was going to be a lot of hand mucking to do to clear the track. "Looks like we'll have a lot of work to do cleaning this mess up," I said to Anthony.

"I guess so" was all he said back to me.

None of Frankie's diatribe upset me much. I was just thrilled to have survived, and, although completely soaked, I otherwise seemed uninjured.

As Frankie had been storming down the drift, I'm sure if he could have fired us he would have, but since we were still alive and the mine was desperate for bodies, we still had jobs. Although we'd survived I always felt certain that had we drowned, Frankie would have fired us posthumously, telling everyone, "You know, I'd have fired those motherfuckers."

Back at the station, Anthony and I called the hoist-man for a man-trip back to the surface, where word of the incident had already preceded us into the office of Mel Vigil, who was awaiting our arrival.

Over a month had passed since I had last been in Mel's office. As I stood there in the doorway, soaked from head to toe, dripping water onto the floor, the sight could have been no less amusing than that first day I had appeared before him, dressed in my new clothes with shiny gear in hand. Sure enough, there on his face, I noticed, was the same bemused look.

I never once considered that either I or Anthony might be fired, and in fact, Mel didn't seem nearly as upset as Frankie had been. All he asked for was an explanation and a few details of the incident.

I told him how, by the use of the powder, we had managed to successfully open the ore chute door. We had been unable to stop the flood once it started, and Anthony and I both had gone underwater. Mel, having undoubtedly seen far bigger knucklehead moves during his career, did not seem particularly upset by any of it and told us to dry off then report to the Dry foreman, who would put us to work. Finally,

Underground and Radioactive

Ore chute door on the main track drift. The top of an ore car is visible directly underneath the door. The pneumatic arm with compressed air hoses attached opened and closed the door and was operated via a small handle to the right that is not visible in this photograph (photograph by R.D. Saunders from an exhibit courtesy New Mexico Mining Museum).

he told us to be ready to go back down, meaning back to work underground, the next morning.

After the incident with the 907 chute, I didn't expect to get many good assignments from Frankie. On the other hand, I had been filling in latrines and hadn't gotten a job that might be considered pleasurable up to that point anyway, so whatever he had in mind for me, I was ready for it.

Thankfully, Frankie didn't hold any grudges, and the following day he greeted Anthony and me both as if nothing had happened. In fact, it was this day that I got my first assignment to work with an actual miner.

Progress

As Frankie approached me in the lunchroom the following morning, I was ready to hear more on the subject of what I had done at the 907 chute and how worthless I was. Instead, I was surprised to find that he seemed to be in a better mood. "Listen," he said, "Schultz is on vacation for a couple of weeks, I put Riordan's helper on opposite shift, so I'm putting you with Al Riordan. Catch a ride on Bordan's motor. He's going back with some stuff for Riordan. Tell him he's giving you a ride. Don't fuck up."

Inwardly ecstatic, I told Frankie I wouldn't fuck up and headed out of the lunchroom door to the dump station looking for motorman Jim Bordan, who would give me a ride down the main drift to the termination heading where Riordan was working.

Incidentally, Oscar Schultz was one of the toughest men I ever saw at Section 35. He was German and had worked in mines all over the world and really knew his craft. A very friendly guy, he was superior at working track drift mining.

A shortcoming of his was that he broke almost every single safety rule there was. He did wear a hard hat, but that was about it.

I once had to deliver some supplies to him at the heading where he was working. As I approached the area, I could tell there was no ventilation. Wherever he had stopped hanging vent tube, it wasn't anywhere close to where he was now. The thing that really got me, though, was he was taking a short break sitting on boxes of powder with a cigarette in his hand. It wasn't the last time I saw a miner doing that, but it scared me something fierce, and I got out of Schultz's heading as fast as I could get rid of the supplies I was delivering.

The miner I was assigned to, Riordan, was the track drift miner working the opposite shift from Schultz. Track drift miners lengthened the main drift by drilling, blasting, and then laying down rail for the motors to run on. It is an arduous job where some of the strongest, most experienced, and technically proficient miners working at Section 35 could be found.

In addition to drilling and blasting unusually large rounds, track drift miners had to lay the ties and twenty-foot sections of steel rail that the motors ran on. Blasting for track drifts required precision drilling so that the floor was flat enough to lay the ties and rail.

Track drift miners installed various types of ground support for the back and ribs of the drift, hung ventilation tube and water and compressed-air pipelines, and operated the largest mucking and drilling machines in the mine.

Miners working track were paid for the number of feet of progress they made each day. A range of $100–$150 an hour was not unusual.

It was good fortune that I would be getting to work on a track drift as a temporary miner's helper. It could lead to other things, perhaps even an assignment as a full-time miner's helper.

One way to quickly get around the mine was to hitch a ride with a motorman as he made his rounds pulling ore or delivering supplies. When I found Bordan by his motor, he told me to hop in a car and he'd take me right to the end of the line where Riordan was.

The ride back to Riordan at the main drift heading, called the face, took about fifteen minutes. Borden stopped the train, and I hopped out of my ore car, thanked him, and started walking toward the end of the track where Riordan was waiting.

I was very happy to find that Riordan had gotten some ventilation into the heading and decided that if I did nothing else, I would be sure that vent tube kept up with our progress.

Getting assigned as a miner's helper on the track drift carried with it a certain level of prestige among the laborers. Some track drift helpers averaged more per hour than a stope miner might make. That was possible because of the contract system that miners working for Kermac were paid under.

Rather than pay an hourly wage, Kermac paid a miner for everything he or she did. Everyone else was on an hourly wage. Each task, no matter how small, had a set price that the miner was paid. Be it setting a single rock bolt, driving a wedge, producing carloads of ore, or constructing timber sets, all the tasks performed during a two-week period were added up and then divided by the number of hours a miner worked to come up with an average hourly contract rate. Miners then had the option of keeping all the contract hours for themselves or giving some of those hours to their helpers.

Contract rate varied from day to day, and the helper, who was usually doing the brunt of the heavy labor, had a big part in this. Some miners were stingy with their contract time and would give only an hour or two per day to their helper, but the good miners always gave the maximum four hours per day.

There was a lot of incentive built into the contract system, because the more you did as a miner, the more you got paid.

Not every mining company in the Ambrosia Lake area used a contract system, with many paying a straight hourly wage. I heard a lot of complaints about other mining companies' hourly wages from some new hires that showed up at Section 35.

I never knew a miner to lie about what he had done, and only once, when a miner reported a $2,500 shift, did I see a shift boss question what a miner accomplished in a day—and that miner was me.

As with many of the most experienced and highly paid miners, Al Riordan was in his prime. Thirty-three years old, he was a rough, sinewy, hardened guy with the heavily muscled forearms and strong, thick hands and fingers I often saw on miners. Dark complexioned with jet-black hair, Al always seemed to need a shave, which gave him an especially rugged appearance long before that look was cultivated by the fashion conscious.

Al was one of the chain smokers at Section 35 who never seemed any more concerned with regulations than he was with blowing himself up, or about the proven connection between radon and lung cancer, about which, we had all been repeatedly warned. The way he always smoked when working with powder scared the crap out of me. In that respect he and Schultz were alike.

I hadn't spoken much to Riordan, but he knew who I was and seemed friendly enough. As I approached him, I had no illusions of getting any contract hours, but this was certainly an opportunity to prove that I could work hard, keep quiet, and follow instructions, all of which I intended to take advantage of.

With a cigarette dangling from the corner of his mouth, he greeted me. "Ready to get to work?"

"Yup."

"Let's get to it then." And with that my real education in mining began.

Attempted Rape

Our second twenty-one-day stay at the Coal Mine campground was nearing an end when the forest rangers paid Greg and me a visit.

As I mentioned, the campground was full every night of the week with more and more transients like Greg and me, which was something the rangers had noticed. We were told in a friendly way that we would have to check out and stay checked out. There would be no turn-around check-in this time. As with many other things that happened up there in the forest, I wasn't sure they could legally do that. It might have been legal, but even if it wasn't, we had little recourse.

We were weighing our options, which more accurately meant only one: become squatters in the surrounding forest. Leaving Coal Mine for a place in the forest where there was no water, no shade, no table, no bathrooms, and no roads wasn't something I was looking forward to after the relative luxury of the campground. Not the best prospect, but it was all we had. As we were contemplating our move, things got worse.

Greg and I both had come off of a four o'clock to midnight swing shift. Dead tired, we crawled into our respective sleeping bags and quickly drifted off. Not long after that, I was awakened by an awful-sounding scream the likes of which I hadn't heard before. At first I thought it might be an animal in the campground but couldn't really make out what kind. I looked over at Greg, who by now was also awake, and asked what he thought it was.

Then we both heard what sounded vaguely like a call for help, so I jumped up and without getting dressed charged out of the tent. Not a good move.

With my first step, my bare right foot landed on a cactus. I might have screamed but thought better of it upon seeing the silhouettes of two people, one holding a pistol and one a shotgun, pointed at each other, in a campsite about a hundred feet away.

By this time Greg was outside the tent too, but he'd had the good sense to put some shoes and pants on, so he didn't suffer my fate.

By either moon or starlight, it was fairly bright that night, so I could see a little of what was happening. I could also hear some of what was being said, but not much. It became clear that the man holding the shotgun was a camper who had also been awakened by the screaming and had confronted the man holding the pistol.

They say if you point a weapon at someone, you'd better be willing to use it, but obviously the holder of the shotgun was not because the man with the pistol raised his arm as if to shoot, and the camper with the shotgun just stood there.

I was expecting to hear a shot at that point, but fortunately there was none. Instead, the man with the pistol began to beat on the shotgun holder using the butt of the pistol.

Most evenings locals from Grants, mostly youngsters, would come up to the campground and cruise through, rarely stopping for anything or bothering anyone. That usually went on all night, even at 1:00 a.m. It so happened that one of those cruisers was coming down the road past our campsite. Greg and I flagged down the car and asked for a ride around the campground. Our hope was to get a license plate number without being discovered. The young high-school-age driver agreed, so we hopped in, me in my underwear and bare feet and Greg dressed properly, and around the campground road we went.

We were unable to discover any information after a single pass, so the cruiser dropped us off again at our campsite. We asked the driver to go to town for help. I figured a good driver could get into town and either to the police station or call for help in about fifteen minutes.

Having absorbed a pistol whipping, the man with the shotgun had recovered, and the two of them were in a standoff again. It was fortunate that to this point nobody had been shot, but we could still see each of them pointing their weapons at each other. I still couldn't catch much of the conversation myself, and neither of us was anxious to get much closer to listen in. We just waited and hoped that the police would arrive.

About a half hour had passed since the cruiser had left for town

Attempted Rape

promising to get help when, sure enough, a couple of sheriff's deputies arrived on the scene. That ended the standoff as one woman and two men were taken into custody, including the pistol and shotgun owners.

More deputies arrived and began questioning campers. When they came to us, we told our story and became witnesses who had to give statements, so we were taken into Grants to do just that.

It turned out a couple of men who had been in town for trial on armed robbery charges had taken time out from that to kidnap a local girl, whom they then brought to the campground to rape. What made it even worse was the girl had obvious mental disabilities. It was she who had been doing the screaming, and why so much of the little we could hear was unintelligible.

Greg and I were identified to the sheriff's deputies by the rescue driver as having been witnesses to the incident. I don't know if there were any other witnesses, but for a campground being as full as it was and for as much noise as there was, it has always interested me that nobody else came forward as a witness.

Greg and I were then asked to follow the deputies into town to help sort out what had happened. We gave our statements to the sheriff and met the girl's father, who was thanking us profusely for getting help.

Neither of us thought we had done much of anything, but unfortunately the two criminal masterminds felt otherwise. The sheriff's deputies let us know that the suspects had made threats against us, and now, knowing some of the accused men's past criminal history, there was no reason to question their word on that. Despite the threats, though, we never saw or heard of the two perpetrators again. Perhaps the robbery trial hadn't gone well and the two men were transferred to more permanent accommodations, but in any event, the whole incident blew over.

We never heard from the camper with the shotgun either, who I would have thought might have shown a little gratitude, but that never happened. Nor did we ever get called as witnesses and never heard anything further about the case—that is, with the exception of the forest rangers.

The day following the incident, the rangers came by and told us that because of our actions, they were not going to evict us from the campground and would instead let us stay another two weeks. That was good news in that we still had a comfortable place to live but bad news because the suspects in the case would know where we were. We weren't giving up all the comforts of the campground despite the threats, so we decided to stick it out.

That incident ruined my working life for a while, seeing as I had a heel full of cactus needles. Persevering in spite of the agonizing pain, I never once considered seeing a doctor, even though Kermac provided 100 percent free medical insurance. I had worked hard to move up a rung on the ladder of the laborer hierarchy and didn't want to lose my place by taking any time off for a doctor's appointment. That was out of the question. And did that heel ever hurt.

I don't know how I managed, but I did. It took quite a while for those needles to work themselves out, but I can attest to the fact that the body can heal itself. I also have a great deal of respect for the cactus and its built in self-defense mechanism, always giving a wide berth to any I see.

A Place to Call Home

Those two extra weeks the rangers had granted us seemed to fly by, and Greg and I were again faced with eviction. The rangers again let us know that we wouldn't be able to check back into the campground after this, so we were out. Whether it was legal or not, we were going to be out this time.

We spent some time scouting the surrounding area before finding a secluded spot on top of a small mesa. That mesa top turned out to be a seemingly perfect spot, as it wasn't far off the main road and was fairly easy to access.

Because we weren't sure how the rangers would feel about our continued residence on their mountain, we thought it prudent not to be seen. As luck would have it, there was a small depression on the mesa where we pitched our cabin tent that made it impossible to see from the main road.

The one drawback to that location was the huge volcanic plug located to the east about a hundred yards away that caused the sun to set very early in the afternoon. We had to bring our own water and had no bathroom and no shower, but we had those things at work, so we thought it was a good spot and one we could live at for quite some time. Not so.

We'd gone on for about two weeks, thinking we had a nice secluded life, when along came the rangers one afternoon. I don't know how they found us, but they did. We were warned that we could no longer live on national forest land and would have to leave. There was not much we could do but agree to that, and we did, but of course we had no intention of leaving because there was no place else to go. So we stayed.

A few days later, Greg and I came back from day shift to find our tent down in a crumpled heap. All the tent poles had been broken, snapped in two. I can't say it was the rangers who had done it, but we had no other suspects, so here again, that's something I'll never know. It could have been a bear with thumbs. We were undeterred by the broken poles, though.

Although Greg thought the tent was useless, I thought I could fix it. So I proceeded to repair each tent pole by crimping the broken ends and forcing them into each other until a couple of hours later the tent was back up.

It might have been a little shaky, but it was standing and looked good. Greg told me there was no way my repairs would hold up, but I declared there was no way they wouldn't. I could see the questioning look on his face, so I further assured Greg that there was no way that tent could be knocked down. I did solid work. Those poles were fixed. Greg went into the tent to arrange his things, and I went back to the car to fetch something.

I couldn't have been gone more than a couple of minutes, but when I returned, the tent was down—all but one corner, that is, through which I saw a face peering out at me through a small screen window. Oops. I will never forget that sight, as it was one of the funniest pictures I have ever seen.

Still undeterred I proceeded to work on the tent poles until I again had the tent standing. I'm proud to say it remained standing for the duration of our stay.

I always expected to have the tent further vandalized, but it never happened. Perhaps whoever damaged it originally thought maybe they had gone too far and just gave up.

We continued to live in our broken tent in the little depression on the mesa for a few weeks until one day a shift boss we knew took pity on us and offered us his cab-over camper to stay in. It was the type designed to fit in the bed of a pickup truck. The question was, where to put it?

Just then providence struck, and a coworker, Ben Wilson, told me he was quitting Kermac and returning home to Butte, Montana. Ben offered me his rented mobile home that was to be available within two weeks.

I accepted immediately of course, sight unseen, and then asked if

A Place to Call Home

Home sweet home, just south of San Rafael, New Mexico. To the rear is the site of Fort Wingate, an Army post established in the 1860s. Kit Carson was brought here to control the Navajos. Arthur McArthur and family, including Douglas, were stationed there for a short time. I would occasionally find an old button from a military jacket while wandering around out back (photograph by R.D. Saunders).

we could park the camper on his rented property in the interim. Ben agreed, and we then had a comfortable place to live and, even better, running water for the first time in a couple of months.

Between and just south of Grants and Milan is the small town of San Rafael. It's just to the west of Route 53 and during my time in the area had a population of around six hundred. The population of San Rafael seemed to be made up of primarily natives of the area who had been there long before uranium was discovered. Maybe some of them worked in the mines, but other than that, there was no visible industry of any kind.

San Rafael appeared to me as somewhat of an oasis in what otherwise was desert country. The main street was almost picturesque with all its greenery, and I would occasionally turn off Route 53 just to drive through the village.

Underground and Radioactive

The place that I was going to rent was a half mile south of San Rafael on Route 53, where there was little greenery but a lot of surrounding wide-open country. It was close to the former site of the original Fort Wingate, where Douglas MacArthur's father, Arthur, was stationed in 1880, and it was there that we placed the camper we would live in for a couple of weeks.

Shortly before moving into the vacated rental home, Greg Hornaday left Grants and his employ at Section 35. I don't know what happened at Section 35 that precipitated his decision to move on, but he decided to go one day, and the next he was gone. It could have been nothing specific but instead an accumulation of things or maybe being just plain tired out.

For the first time since arriving in the Grants area, I had an actual home now and one that I thoroughly enjoyed. It wasn't close to any of the crowds in Grants or Milan, nor was it anything fancy, but that was fine with me. Only some ranchers and a few miners lived out there, so despite being barren country, it was peaceful and offered some great views of Mount Taylor.

Miner's Helper

I continued working with Al Riordan for two weeks and enjoyed it the entire time. He was a very good miner and a great guy to learn from.

It was the first time I saw drilling and blasting up close, and while I didn't know exactly what Al was doing, I could see *he* certainly knew what he was doing. Each round he blasted seemed to turn out perfectly, always the same height and width, as we advanced the main drift. I still don't know how he was able to get the floor so flat so that the ties and rails we laid were perfectly level.

I learned about ground support, both about the temporary roof jack and more about the permanent stull. I had earlier loaded and unloaded a lot of stulls, so it was good to see how they were used. I learned how to scale the back and ribs using a scaling bar. A scaling bar was a long six- or eight-foot piece of heavy steel with a flattened end by which any loose material is brought down by prying, before it fell on its own or on us.

One of my main jobs during track drift construction was to scale the back and ribs after every round was blasted. My hard hat always got a lot of use when I was scaling, as I managed to pry rock onto my head from time to time.

When the ribs were sufficiently scaled of loose material, Al and I would cover them with steel fencing material, fastening it with rock bolts.

Track drift miners used a mucking machine to clear the blasted material away from the face and into waiting ore cars for disposal.

Mucking machines were somewhat like little front-end loaders that operate on the rail in the main drift. They had a small bucket on the front that was used to scoop up the muck from a blast. The mucker

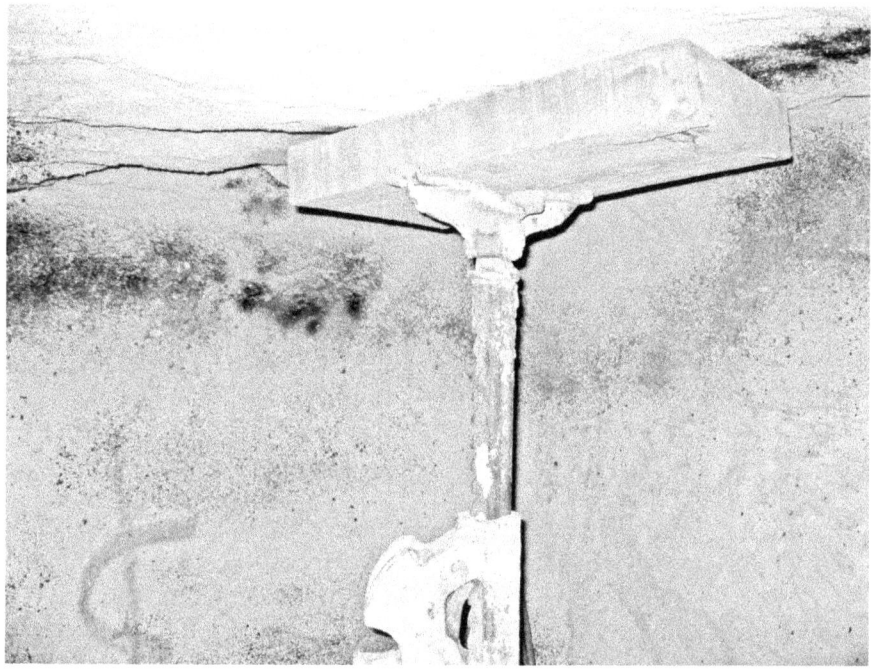

Roof jack in action. We used these to support the ground directly above our heads when drilling. They were easy to use and very effective (photograph by R.D. Saunders from an exhibit courtesy New Mexico Mining Museum).

operator, standing on a small platform attached to the side of the machine and using hand control levers, would flip this hinged bucket full of muck over his head, throwing the materiel into a waiting ore car behind the mucker.

As with most large machinery underground, the mucker was run by compressed air and was a lot of fun to operate if you knew what you were doing. For an inexperienced operator like me, a mucking machine was very easy to derail. Al let me run the mucker a few times, but I derailed it, and that cost us time and therefore money. I did, however, learn how to get a derailed mucker back on track very quickly and efficiently.

Now and then when Al would take a break or go back to the station for something, I would practice running the mucker up and

Stulls in action. Set precisely vertical, they provided excellent ground support for relatively small areas (photograph by R.D. Saunders from an exhibit courtesy New Mexico Mining Museum).

down the track, getting familiar with the controls. That came in handy later when I had to use a mucker full time on another assignment.

Hauling rock from the face of a track drift was another one of those jobs that a motorman didn't want to do. Only hauling ore paid a decent contract rate, and while hauling waste material paid something to the motorman, it was hardly worth his time, and unless under direct orders from a shift boss, a motorman was hard to find when waste needed to be hauled.

Because track drift mining produces waste muck almost exclusively, it's left to the track drift miner's helper to haul the waste material. Thanks to my previous experience running motors, it turned out I was the guy for the job. So when Al asked, "You ever run a motor?" I was able to respond, "Yup." So Al said, "OK, go get a motor and get back here; we have a lot of work to do." One thing about working for Al was he never let up. It was go, go, go all the time.

I went all the way back to the station, this time walking, picked up a motor and an ore car, and drove back to the face where Al was busy drilling another round.

Drill, blast, scale, muck, haul the waste, set the ties, wire the ribs, and lay the rail. That was pretty much the job as a track drift miner. Now and then we would also erect three-piece timber sets, consisting of two posts and one cap wedged against the back for enhanced ground support.

Moving and installing the long lengths of rail was extremely difficult. The rail was unwieldy, and again the steel-toed boots came in handy a few times as I would drop rail ends on myself now and then. With those steel toes, the rail would just bounce off.

As difficult as it was to lay rail, it was only a narrow-gauge track, so it gave me some appreciation for what it must be like to install full-size rail.

It was extremely heavy lifting working track drift; there was not much danger, and it was otherwise boring, but man did it pay. Al was making about $150 an hour, which I knew before I ever worked a day with him because each miner's contract rate was posted in the Dry every other Friday.

I think he might have thrown me an hour or two of contract those

Miner's Helper

two weeks I was with him, but I hadn't expected any at all, being just a fill-in guy, so I was grateful for anything.

I was most appreciative of the temporary promotion to helper and the opportunity to finally learn something about mining. And this led to my next job.

Down the Chute

Manuel Bustos was an excellent, experienced miner in the prime of his working life. As had been the case with Al Riordan, Manuel's regular helper was off for a couple of weeks. Bustos was cheerful and friendly, even around the new hands, so when Frankie said to me, "You're with Bustos today," I was pretty happy about it. It must have meant I did all right during my time with Riordan.

Bustos was a, strong, stocky man with a full, scraggly beard and longish, jet-black hair. He had kind of an odd look about him with no butt to speak of, which was only noticeable because of the extra fabric that dropped down low as he walked, causing him to be constantly pulling his pants up. He always wore his miner's boots over his pant legs, which I found to be a particularly funny look on him. He had a great sense of humor and some interesting stories he'd accumulated through the years.

Everyone carried a steel thermos bottle underground along with their lunch bucket, and I'd noticed that Bustos's thermos had a very large dent in it, almost separating it into two pieces. Because he was such a friendly guy, I asked him about the thermos one day during lunch break.

For a while early in his career, Bustos had been a surface worker. I think he said he ran a loader, but one day he and his crew sat down next to a loader to have their lunch. Bustos placed his thermos on the ground next to him when suddenly the loader bucket that had been left suspended in the air ten or twelve feet let loose and dropped right next to the crew and on top of Manuel's thermos, bending it practically in two. He'd saved that thermos—I guess as a reminder to always be alert to his surroundings.

Bustos worked the pillar stopes, so named for the pillars of rock left in place to hold the ground above from collapsing after an area had

Down the Chute

Example of a pillar stope. During the mining process pillars of ore are left in place as ground support. To ensure all the ore is recovered the pillars are later removed and the ground above supported using other techniques (United States Department of Energy archives).

been mined. Pillar stope miners produced a lot of ore and were very highly paid as a result.

Motormen loved the pillar stope guys because of the large volume of ore they produced and would sometimes pull the ore from one stope all day long. It was fast, it was easy, and they made a lot of money doing it. But it was a two-way street.

As a pillar stope miner, it paid to be on good terms with the motormen. Motormen had plenty of other stopes to pull, so if a miner gave them a hard time, it could be a while before a motor came around to pick up ore. Those miners who had a problem with a particular motorman could find their ore sitting in the chute and production stopping for a long while, and down would go the miner's paycheck. That's what happened to Bustos.

Jerry Sanchez was the lead motorman at Section 35, and his brother, Leo, was a hoistman. One morning Leo let the cage that Bustos was riding free-fall a couple of hundred feet. That's a serious drop to be on the receiving end of.

Manuel was still hot about it after his shift ended and went looking for Leo, and while no fight ensued, there was some bad blood between the two that spilled over to Jerry.

Bustos probably should have known that would happen, but if he did, he didn't care. The end result was that Jerry Sanchez found every reason not to pull the ore that Bustos was producing. Sure, Bustos could have gone complaining about it and gotten some ore pulled, but that hurt some other miner, so he saw no sense in it.

Like most good miners, Manuel Bustos was not one to sit around waiting. As the ore in Bustos's stope filled the chute, something would have to be done if he couldn't get Jerry to pull it. There was only one thing to do in that case: pull your own ore.

While Bustos was an excellent miner, he proved to be a lousy motorman and would have been better off having me do it for him. But I guess he was stubborn and too emotionally invested for that, and he would show Jerry Sanchez who the real motorman was.

The uranium ore collection site was located just off the station. To the back of the cage was another lifting area in the main shaft where the ore hoppers were. As long as there was ore in the collection chute, those hoppers were running up to the surface.

Motormen would pull their trains into the collection area, lining the cars up with one of two very large gates. The gates opened inwardly so that the ore could be dumped into collection chutes that were much bigger versions of the chutes at the stopes.

The side of the ore car facing the chute was hinged. On the reverse side was a clevis to which a cable was attached that was also connected to a winch. The motorman first pulled a handle located to the side of the gate, opening it, and then, using a separate handle, lifted the ore car on its hinges, dumping the ore into the chute. While it sounds involved, a good motorman was very fast and could unload ten cars of ore this way in under five minutes.

The tricky part was getting the ore car to flip back over and drop down onto its chassis. That was accomplished by pulling the car over

Down the Chute

fairly quickly via the winch so that the car dumped its ore and the weight and momentum of this stretched the cable slightly so that it acted like a steel rubber band. The bungee effect would bounce the car back to the tipping point, where it would then drop back over and be lowered by the operator to its resting position. An experienced operator completed this process very quickly, and the only people in the mine with that kind of experience were the motormen.

It was fun to watch an experienced motorman dumping ore, and although I tried to master the technique, I never got the hang of it. It usually took me several tries to get the ore car back over and settled on its chassis.

The winch and gate levers looked identical, and perhaps at one time someone had thought to color code them, but if that had ever happened, the color was long gone. I looked at those levers and never could tell the difference myself so was always careful around the ore dump. But there was no real need for an experienced motorman to be able to identify which was which; they just knew. Not so Manuel Bustos.

Always a man in a hurry, Bustos could be seen all but running from place to place. Regrettably, he applied his otherwise enviable work ethic every time he tried to be a motorman. Forgetting to turn on his trip light, speeding, and failing to pull the safety light lanyards hanging from the back were all his trademarks.

It didn't hurt to know when he'd be pulling ore, as it was an invitation to be even more vigilant while walking down the track drift. He pulled a lot of ore this way, but it came close to being his undoing.

Much as I'd done working the main track drift, I did my best to help Manuel. An important component of that was keeping my mouth shut. I didn't know anything, and unless I saw something falling, I wasn't about to say anything, let alone question what he was doing. So when he told me he was going to start pulling his own ore, I certainly wasn't going suggest I do it myself. If he had wanted me to do it, he would have said so, and besides there was the matter of his pesky emotional attachment to the problem.

As we worked in the stope, there came to be many tons of ore backed up. I don't know the exact number, but the hundred-foot chute was full, and we had an awful lot of ore lying around. Jerry Sanchez had enough to keep himself busy and was not going back to pull all that ore, of course, so Bustos took matters into his own hands.

Bustos collected a motor and enough cars for a twelve-car train and made his first trip back to the stope's ore chute. As I operated the ore chute lever filling the cars, Bustos would pull the motor ahead. Here was where my previous ore chute opening experience really made a difference, for in no time we had those cars full and together made it to the dumping station uneventfully—although I have to say I'd never ridden in a motor at that almost reckless speed.

At this point I became a spectator. Still, in a hurry as always, Bustos jumped from the motor heading for the winch cable handle to get some slack to hook up his first ore car, hastily yanking on the lever. Regrettably, not only did he pull on the wrong handle, but he was also standing on the lip of the gate, which opened as quickly as he had yanked on the lever. The speed at which the gate opened gave him no chance to catch himself, and Manuel Bustos had dumped himself down the ore collection chute—potentially a very long drop.

Those ore collection chutes were very deep, and a drop like that could be fatal; otherwise, the sight of Manuel disappearing down the chute would have been very funny. That chute flew open and down he went, feet flying up in the air and hands flailing as he quickly disappeared. All that remained to be seen was what was left of Manuel Bustos at the bottom of the hopper shaft.

If the chute was empty, or even a third full, it would mean one long fall onto a pile of rock and in all likelihood a fatal one. That's what I was expecting as I inched up to the open gate to have a look.

Manuel Bustos, who had shown me before that he had some luck, had fallen only twelve feet, as the chute was almost full. He sustained no injury other than that to his pride and was able to crawl out.

Of course, as I was the only witness, there was no chance I would ever repeat what I had seen unless to confirm whatever Manuel decided to tell of it. He, being a good-natured man, actually delighted in telling the story, and I ended up having to confirm that what he said happened really did happen.

We pulled ore from the stope all day after Manuel's close call. It had to have been more than a hundred cars—a good payday for Bustos and a good story for me.

The Iron Blossom

I had never been much for saloons. Not that I had anything against a drink now and then, but it just wasn't my style to hang around bars. However, I soon found that if I wanted to socialize with my new mining buddies, then I needed to know something about the local bar scene.

The first thing I learned about the saloons in Grants was which establishments were relatively safe to go into, and those that were not. If the consensus I heard was to stay away from a particular bar, there was always a good reason or two. Whether it was because of a stabbing, a shooting, or an excess in fighting, I would stay clear.

On the other hand, some saloons had good reputations as places miners could go and, relatively speaking, safely have a good time. The Iron Blossom Saloon was one such place.

The Iron Blossom was a popular spot, and most of the people I associated with could be found there. It had large, round tables with eight to ten chairs and probably fifteen tables total plus a long bar. The long bar had been imported from a saloon in Butte, Montana, and actually was a beautiful piece of furniture in its own right backed up by a large mirror. The Iron Blossom setup was very similar to saloon sets seen in television and movie Westerns.

Attached to and behind the Iron Blossom was the La Ventana restaurant, which is still open today. It was a very comfortable place for a meal back then and still is.

I can't remember a time when there wasn't a sheriff's deputy stationed in one corner of the Iron Blossom. The story I heard at the time was if a bar didn't have a deputy, it couldn't be open. I don't know if that was true, but I know what I saw. You would think with a rule like that the place would be expecting trouble, and sure enough, I saw some interesting fights in the Iron Blossom.

Underground and Radioactive

One night we had a group of ten sitting around one of the large, circular tables. I believe all of us were drinking shots of Wild Turkey that night; at least I know I was. As time passed and, with things getting a little blurry for me, a fight broke out somewhere in the bar. Before long a body came hurtling through the air right into the middle of our table. Whoever threw that person through the air had to have been one strong guy, because he was really moving when his body hit the table.

The next thing I know, the table had moved several feet away, and the people who had been sitting around it scattered in all directions. Except me. I was left sitting there, having never been touched, in a chair by myself, with no table and no other people. Just me sitting by myself. It felt strange, and it must have looked even more so.

The fight was short lived, and order was soon restored. The table was put back in place, although not by me, I can assure you, as I was in no condition to move anything. Soon my friends returned from having been tossed about, and our happy evening continued.

I only ever came close to having real trouble at the Iron Blossom once—at least that I'm aware of. A few friends and I were taking up about half the chairs at one of the tables. Around the other half were three or four guys I didn't know. That happened a lot, so it wasn't unusual at all. My friends, who were all from Butte, and I were talking about Montana and making jokes about it. I don't recall any of what the jokes and comments were, but they really riled up a man named Joe sitting across from us.

Joe was unpleasant to be around, acted surly, and my guess was he had a short fuse. I hadn't run into anyone like him in Grants before, and he didn't seem to fit in. He might have been someone who liked to hang around boomtowns but not actually work much.

When Joe got up to use the bathroom, his friend came over to me and told me how upset Joe was that we were joking about Montana. "Oh, we're just fooling around," I said.

"Yeah, well, Joe here doesn't like it, and I'd knock off the jokes if I were you. You don't want Joe mad," he said to me.

"What, he's from Montana, I guess?" I replied.

"He just got out of Deer Lodge," he said. I wouldn't mess with him."

"I don't know where Deer Lodge is, but no problem; we're all just having a good time. Nothing personal."

"Deer Lodge is where the Montana State Prison is, and Joe just got out a few days ago."

Armed with that important information, we said no more about Montana the rest of the night. Even so I could see Joe was fuming and ready to start trouble.

As the hours went by, I bought a few rounds of drinks for everyone in hopes of mollifying Joe, who continued to sit at our table. I could see that buying him drinks wasn't working very well, so I needed a better plan. If I had gotten up to leave, there is no doubt that Joe would have followed me out and a fight would have ensued. I don't know what the end result would have been beyond not good for anyone.

Being familiar with the Iron Blossom, I knew there was more than one way out that Joe, being new to the area, wouldn't know about. As luck would have it, the second exit was just beyond the bathrooms through La Ventana. The bar closed at 2:00 a.m., so at around 1:30 a.m., I got up, ostensibly to use the bathroom. I walked through La Ventana and out the door to my car.

I never saw or heard of Joe again and have little doubt he ended up right back in Deer Lodge or some similar institution.

You put a bunch of miners with a lot of money to spend in a bar, and things happen. I was quite content to sit, have a drink or two, listen to music, and tell stories of the latest happenings at Section 35. On the other hand, some miners weren't, and things could get fairly rowdy from time to time. I have to hand it to the Iron Blossom, though, because the bouncers they employed contributed to it usually being a relatively safe saloon compared to some other places in town.

The other thing that caused some real problems in the bars was the relative lack of women. Very few women worked at the mines. Some did but they were few. I knew of two at Section 35 who were underground, and there could have been some on the surface, but I never saw them if they were there. A few single women were in the area working in retail and at the bars and restaurants, and of course quite a few miners brought their wives. There being very few single women around, made them very popular among the male mining population.

Based solely on my own observations, I'm not sure a miner bringing his wife to a boomtown like Grants was such a good idea. I'm certain there were wives and husbands who had no problem and got along just fine, but I didn't see many. That might well speak more about my circle of friends than anything else. Nonetheless, there were some humdinger fights over wives and single women.

There were quite a few good restaurants in town besides La Ventana, including the Grants Cafe, the Monte Carlo, the Holiday Inn, and the Uranium Cafe. I wasn't a really good cook myself, so I spent some time in all those places.

The fast-food chains consisted of a Tastee Freeze, a Pizza Hut, and some independent hamburger spots. If you wanted McDonald's, it was sixty-mile drive to Gallup, something I did more than a few times.

Revenge

Don't get blown up. It sounds easy enough, but there are a variety of ways to do it underground. I found that the right amount of experience and following safety rules greatly reduced the chance of being caught in a blast.

Kermac did what they could to help eliminate explosives accidents by using electric blasting caps and centralized detonation of all rounds as a rule. That assumed miners followed the rules, and many did not.

I worked with a few miners who stayed within the blasting rules that Kermac had established and several who did not. I can say that when my turn came, I often skirted the rules myself and am fortunate that neither I nor anyone else was ever injured as a result. I'm confident that what I did to get the ore chute open using powder was a violation of safety rules, and I was very lucky not to have been injured by blasting that way.

It wasn't uncommon in my wanderings as a laborer to find a miner sitting on a powder box puffing away on a cigarette, as had been the case with Al Riordan when I worked with him on the track drift. While he never blew himself up that way, I think the relative stability of powder would have been compromised had it been exposed to a direct heat source like a cigarette.

I didn't care how stable the sticks of powder were, I wasn't messing with them. Johnson and Jones might be throwing sticks of it at each other, but I always looked at powder the same way as that supposedly unbreakable china sold in discount stores. The salesman says, "Watch this," and drops the stuff on the floor, and the unbreakable china shatters. Then the salesman says, "I've never seen that before." I could always imagine Johnson and Jones throwing powder at each other and suddenly Jones vanishes in a pink mist. Johnson says, "I've never seen that before."

It does, however, take a lot to set off a stick of powder under normal circumstances. In a planned explosion, it requires blasting caps to ignite powder. But as I found with my next assignment, sometimes working around powder isn't normal.

After my two weeks with Bustos were up, I walked into the lunchroom the following Monday wondering what I would be doing. The actual jobs a laborer was assigned were never the worst part. The worst part was always wondering what I would be doing from day to day, making those two-week stints helping miners like Riordan and Bustos seem like vacations from the constant uncertainty.

Frankie walked up to me and said, "I want you and Gonzalez to go around to some stopes and collect anything left in the old powder boxes." Handing me a piece of paper he said, "Here's a list of the stopes."

"OK," I said. No problem; I can do that. Easy job.

Then as he turned away, he added, "Oh yeah, some of the powder might have been there awhile. There might be liquid nitroglycerin in the box."

That got my attention. "Nitro?" I said.

"Yeah, the powder gets old and the nitro leaks out. You gotta be careful. That stuff could blow if you don't know what you're doing."

Wait a minute now, I thought, I actually *don't* know what I'm doing, Frankie.

Sensing my concern, he said, "Any nitro in there, just fill the box with water. It'll be OK."

I was to report how much powder we had found, and I was hoping to make that report in person and not by way of a muffled explosion. With that Al Gonzalez and I were on or way.

In taking on this job, I discovered that our mine needed a better system of accounting for sticks of unused powder, because there seemed to be a lot of it lying around.

We spent the shift wandering around stopes looking for powder boxes and surplus sticks of it that might be inside. Sometimes we would run into a shift boss who would let us know where an abandoned box was and, if absolutely necessary, lead my partner and me to it, point at it, and then retreat. When that happened, I was convinced what we were doing was a job given only to nonessential personnel.

The first few boxes we found did have powder but still in stable stick form, so we collected that and started storing it in a cutout on the main track drift.

Soon, however, we came upon a box that appeared to be quite old, and, opening the lid, we saw some paper and sawdust floating in an inch-high pool of nitroglycerin.

Some of the powder boxes around the mine were really very small, holding about a dozen sticks or so. I don't know what they were used for, but in a production stope, a lot more than a dozen sticks would be needed for each round, so most of the boxes we encountered were fairly large, holding several dozen sticks. It was in one of those boxes that we had encountered the liquid nitroglycerin.

Frankie had instructed that if we came upon a scenario such as this, we should fill the box with water to neutralize the nitro. I wasn't sure if that method would work, but I didn't know any better way and was determined to fill that box with water as instructed.

I sent Al back to the station for a water hose, which we then connected to a water line and valve connection on the main track drift. Then we hauled the hose up the seventy-five-foot manway.

I slowly snuck up on the old powder box so as not to cause any disturbance that might jostle the nitroglycerin in it and very carefully opened the lid but only partially. I stuck the end of the hose in there just a little, then carefully lowered the lid, which seemed to hold the hose in place. Because the valve was at track level, I reasoned we would be far enough away that even if the force of the water jostling the nitroglycerin was enough to ignite it, we would be safe.

I climbed back down the manway to where Al was waiting for me and slowly turned on the water. Hearing no explosion, we left the water on for at least ten minutes, or long enough to easily fill that old powder box.

I turned off the water then climbed up the manway to see if the plan had worked. Indeed, it had worked like a charm. The nitroglycerin had apparently been diluted enough to no longer pose a problem. We continued on that day, working our way from stope to stope, sometimes finding more nitroglycerin and then using the same method to eliminate the danger.

We were working in a lot of areas of the mine that had been either worked out and had no more ore or for one reason or another had been abandoned. Principally, these were places where nobody thought anybody would be. I know Jim Peters didn't think we would be where we were.

Jim Peters was an exceptionally experienced miner approaching the age of forty. Someone who had made it to age forty working underground at that time was an unusual sight to see, and although I would know shift bosses forty and older, I didn't see many miners that age. Even as an inexperienced laborer, I picked up on it. Part of the reason was people just wore out physically, and part of it was injuries.

Peters was not only experienced but an exceptionally skilled miner at the peak of his earning potential. Another of the keys to his productivity and common among the best producers was that he ignored most safety rules. He smoked around powder, he couldn't have cared less about ventilation in his stope, he usually worked under unsupported ground, and he blasted whenever he needed to by hand lighting rounds. The guy violated almost every rule in the book, but he really made great money.

Hand-lit rounds were a serious safety hazard in the mine and a major safety violation. The rule was to drill and load your round, then hook it to an electrical box. The wiring from the box in every stope ran to the lunchroom. During lunch break, and at the end of a shift when everyone had been accounted for, the shift boss threw a switch, and all the rounds went off more or less simultaneously. Generally, what was heard in the lunchroom when the rounds went off were muffled rumblings.

As some miners ignored the rule, we heard rumblings throughout the day, but if you were a shift boss trying to determine the location of the hand-lit round, it would have been very difficult. You almost had to be there at the source.

The reason miners didn't use the central blasting systems was money. Time spent sitting around waiting for someone else to blast a round cost the miner a great deal, so many guys frequently hand lit rounds. It wasn't as if the miners who hand lit rounds were reckless about it. On the contrary, they were usually very careful to look around their work areas to be sure no other workers or bosses were close by.

I'm not so sure how much miners cared where the shift bosses were, though, further explaining why the experienced shift bosses tended to stay around the station and lunchroom. No sense getting blown up.

When the miner was satisfied the surrounding area was clear, he would go ahead and light the round.

The lead time on a hand-lit round could be up to five minutes or more, giving the miner and his helper enough time to get as far away as possible but not too long so someone might inadvertently run into a round going off. If someone heard or felt the round go off, he would have no idea where it came from—that is, unless you, as the hapless, inexperienced laborer, happened to be in the wrong place at the wrong time and where nobody expected anybody to be.

As time went by, I'd heard things about certain miners, and by this time I knew of Peters and his reputation and where he worked. I knew that Al and I were in the vicinity of Peters's stope as we searched for old powder.

As we wandered through old drifts, I could hear the far off sounds of drilling and assumed it was Peters. Al and I had been investigating a stope very near where Peters and his helper were, but, finding no powder, we headed back down to the track drift.

I sent Al back to the station for something or other, and while he was gone I figured I would go up to Peters's stope and see some drilling, which was of great interest to me by that time. I wanted to know more about how it was done, maybe get an idea of how to drill and set up a pattern. A good miner like Peters wouldn't mind me watching.

When I found the manway to Peters's stope, it was at least a 170-footer. They didn't get a lot higher than that at Section 35. I started my long climb.

When I came to the top of the manway, I didn't see anyone and didn't hear any drilling. I thought they must be done and maybe they were loading their round. It never occurred to me that as I was climbing up the manway, there was a hand-lit round ignited and hissing up in the stope.

Popping my head out the top of the manway, I paused to listen, and as I did so, the round went off.

Happily, the manway was some distance from the face where the blast occurred, so I wasn't obliterated on the spot, but instead I quickly became engulfed in thick, white, blinding, choking smoke. That was a very bad thing because in an enclosed area like a mine, the smoke had nowhere to go until it hit one of the exhaust fans on the main track drift that were set at various intervals, and I certainly didn't know where that was.

Still standing in the manway entrance, I proceeded to attempt and probably did set a new manway descent record. When I hit the track level, I was still engulfed in very thick smoke, and it was becoming difficult to breathe. I couldn't see a thing for the smoke, but I could hear a fan.

Not knowing where the fan was, I nonetheless felt the direction that the air was moving, so I headed the same way. By now my lungs were hurting, and I was feeling worse by the second. Soon I started thinking about breaking out my rescue unit from its case attached to my belt. This was the exact situation it was designed for.

We had been instructed in orientation that were we to get stuck in smoke, we should break out the air rescue unit and breathe through it, giving us about fifteen extra minutes to figure something out.

Because the smoke was so thick, and I was feeling so bad, I did finally break out my self-rescue unit and was just about to put in the mouthpiece when I came to an exhaust fan, and just that fast the air cleared. On one side of the fan, there was thick, white smoke, and on the other side it was completely clear.

My lungs were burning, and I was feeling mighty sick, but I was going to survive another close call. Then I thought of Peters.

The immediate threat having passed, I started getting angry in a hurry. I could have gone to the shift boss and complained about it, but that would have been seen in the Dry as being as bad as a jailhouse snitch.

One possibility would have been to attack Peters, but considering his experience and pay level, that would have been career suicide, and I might well have lost that fight anyway. No, the best solution to my having been wronged was revenge, and I intended to extract it.

My lungs were really making it painful to breathe, but gradually during the long walk back to the station I began to recover. The walk gave me plenty of time to think about ways I might exact my revenge, and by the time I found Peters and his partner standing by the cage door, I already knew what I was going to do.

I let Peters know what happened, and he did apologize, assuring me that both he and his partner had searched the area prior to lighting the round, and he didn't see how they could have missed me. Coincidentally, I had been wondering the same thing.

Not only had they not seen me, I hadn't seen them. On top of that, I was wondering where my partner was. I had sent him back to the station for some forgotten piece of equipment, and he had never returned. I didn't know what Al had been doing, but when I found him, he swore he hadn't seen anything unusual. I can't say I was surprised.

I do recall being impressed that a miner with Peters's experience would even acknowledge let alone apologize to a laborer. That was nice, but I still planned on teaching him a lesson about hand lighting rounds.

I was feeling subpar for several days as a result of breathing in all that smoke but again never considered visiting a doctor. I would have missed work and might have had to explain what had happened, which could have gotten back to Shotgun, Mel, or Frankie. I didn't want it getting out what Peters had done. One day I hoped to make the kind of money Peters made, so I might be having to hand light rounds myself. I resolved if that were to be the case, I would be a lot more careful about it.

Finally recovering enough to return to normal after a few days, I put my plan for revenge into motion.

Nothing was going to stop a good miner from doing all he could to make money. A little incident like getting a laborer whom nobody cares about caught in some smoke from a blast wasn't going to mean a thing. Peters was going to continue to bend and break every rule if it meant a bigger check. The guys like Peters just kept on driving full speed ahead, which meant he was going to continue to hand light rounds. For my plan to work, that was exactly what I needed to happen.

Through some good fortune, I knew Peters's helper, who felt genuine remorse about what had happened. After I explained what I wanted to do, I convinced him to assist me. His condition for doing so was that Peters was never to know. I assured him all I needed was a heads-up when Peters was going to hand light a round.

I knew it wouldn't be easy to get away from whatever work I was doing to put my plan into action each time Peters hand lit a round, but I knew too that there would come a time when I could. So I waited.

My break came one afternoon only a week later when Peters's helper told me during lunch break that they were just finishing up loading a round and would be blasting shortly after lunch. That day it so happened that I was working close enough to Peters that I could stop what I was doing and head over to his stope.

At about the time Peters's helper had estimated they would be ready to light off a round, I found a cutout relatively close to their stope and hid there. I knew that after the round was lit, the two of them would be walking by in a hurry. I hoped not to be seen by turning off my headlamp, and I waited—just me and my extra pair of miner's boots.

It wasn't long before Peters and his helper came down the manway and walked along the main track drift right past my position. They were in no particular rush, so I figured they had at least a five-minute lead time on their round. Having learned from experience how exhaust fans work, I had selected a hiding spot on the correct side of one and wouldn't be getting engulfed in smoke when the round went off.

I continued to wait there, and sure enough, within a few minutes the round ignited. The smoke poured down the manway and out the exhaust. It was fascinating to observe how fast that smoke moved and how quickly it cleared the mine.

Within just a few minutes, all was clear. That's when I bolted up the manway. I'd already set the record for descending a manway; now I was trying to set a new ascension record. It wasn't easy, but by ignoring all manway safety rules, I think I might have set that record.

On the way up a manway, there are trap doors made of steel mesh every twenty feet or so. Going through one, you would close it so in the event of a fall, it wouldn't be a long fall. To save time, especially in a tall 170-foot manway such as I was now climbing, the trap doors all stayed open.

When I hit the top, I ran to the face where the round had gone off, and there remained a very large muck pile. I took my extra set of mining boots and covered them in the muck pile with just the feet sticking out so it looked like someone was buried.

I then lit out for the manway, attempting to get to the bottom quickly and back to my hiding place. I didn't want Peters to see me as he and his helper returned to their stope.

I made it to track level and went back into hiding in the cutout. It wasn't long before Peters, with his helper in tow, came ambling down the drift and back up the manway. Leaving my hideout, I took off back toward the station as fast as I could.

Peters's helper knew what was up there, but of course Peters had no idea. I wish I could have been there to see it, but I understand there was no small amount of panic when Peters saw those boot heels sticking out of the muck pile.

As fast as I thought I was climbing down a manway, I should have known there was no way I could beat a miner with twenty years of experience. If I had the descent record, it was short lived, because I later heard Peters came flying out of that manway and racing down the drift on his way back to the lunchroom at record speed. It must have been one of the very few times Peters ever needed or wanted to see a shift boss, but there was no way I could let that happen.

While I saw my revenge prank as funny, no shift boss or superintendent would have, so I had to intercept Peters and his helper prior to their reaching the station. So I waited several hundred feet away.

Seeing the two headlamps approaching, I began walking back toward them. When I saw Peters, he was in full panic mode, no doubt wondering how he would explain what he had done.

"What's going on?" I asked, sounding concerned. Peters kept on hurriedly walking past me, but his helper said something about an accident in the stope to which I replied, "Oh, the boots in the muck pile?"

Peters slowed, then stopped, and, turning back toward me, said, "How the fuck do you know that?" Right then the light went on. "Oh, you motherfucker."

"Yup" was all I said.

From there Peters put the whole thing together quickly. While he never laughed about it, he wasn't angry either. He always respected me after that, and, in his eyes anyway, I'd moved up in the Section 35 hierarchy. I got immense pleasure out of my revenge caper for years.

The Sandman

Shortly after my prank, I was assigned a new shift boss, Bill Clark. Another transplant from Butte, he was an affable, low-key man and generally a matter-of-fact kind of guy. I liked Bill right away.

He was a neighbor too. Shortly after finally making my home in San Rafael, Bill moved his family into a large mobile home about a hundred yards away from mine, so I spoke to him from time to time.

My first day under Bill, I was assigned what turned out to be my final job as a laborer, that of sand-fill operator. When a stope has been worked out of all its ore, what's left is one big hole in the ground, and no matter how much ground support is installed, the ground above is going to cave in eventually. It's one of those cases of when, not if. Were the ground to cave in all the way to the surface, that would be a lot of sinkholes around the Ambrosia Lake mining district.

What mining engineers came up with was a system of filling old stope cavities with sand. It would have been next to impossible to fill an empty stope with sand by shoveling it in, so a mixture of sand and water was pumped from the surface and into the stope. There was a massive pile of sand on the surface that was combined with water in a plant nearby, then pumped underground through a network of eight-inch pipes running throughout the mine. The sand would sink right away, and the water above would drain out. In this way it was possible to fill a stope to within an inch or two of the back, greatly reducing or eliminating the likelihood of a sinkhole developing.

From firsthand experience, I learned that sand-fill was a good system that worked well. As I was to find out later, though, before a stope was filled with sand, somebody had better be certain all the ore was out of there, because having to go back in is very difficult and dangerous.

The Sandman

On my first day, Bill told me and another laborer, Daniel Ortiz, to start hanging sand-fill pipe from a junction close to the station on back to the 502 stope. There was usually a large stack of the twelve-foot-long sections of pipe close to the station along with hundreds of rubber-lined pipe clamps. Having worked unloading and stacking sand-fill pipe from my time at the station, I knew how heavy the pipe was but had never hung any before.

It wasn't difficult to figure out how to hang and fasten the pipe, but lifting those sections into place, sometimes eight feet above the floor of the track drift, was difficult, requiring a great deal of strength. I'm tall, but my partner, Daniel, was considerably shorter, adding to the difficulty. In addition, we couldn't very well block the main drift with lengths of sand-fill pipe, so each time we hung a section, we had to walk back to the station, pick up a section of pipe, and carry it back by hand.

Our good fortune was that the 501 had been one of the first ore-producing stopes in Section 35 and therefore also one of the closest to the station at around three hundred yards or so. I would come to know the 502 and its neighbor, the 502 stope, very well during my time underground.

Daniel and I had three hundred yards of sand-fill pipe to hang, so we got after it. Other than the physical labor involved, which was arduous, there wasn't much to it. We did have to run some sections of pipe vertically through the manway and then finished by attaching a final length of it that we hung from the back, which could be swung from side to side to direct the flow of sand-fill. We managed to complete the whole job in a week.

At the top of the manway, we had to build a bulkhead to hold back the sand and water as it was being pumped into the stope. The bulkhead was constructed about ten feet from the manway and was about eight or nine feet tall and six feet wide.

I reported to Bill that the stope was ready to be filled with sand, expecting then to be assigned another interesting job. Instead Bill told me I was to be chief sand-fill operator.

A telephone had been installed by the mine electrician at the top of the manway next to the bulkhead, and I was to man the phone, communicating with the surface sand-fill plant from time to time to report

Sand-fill telephone. I had a lot of experience using this to talk with the sand-fill plant operators on the surface (photograph by R.D. Saunders from an exhibit courtesy New Mexico Mining Museum).

on sand-fill progress. Otherwise I was just sitting there watching sand flow into the stope for hour after hour. Additionally, I was to do it alone. I guess the bosses figured it was so easy and so close to the station that nothing could go wrong that might result in an injury.

At first it was kind of interesting watching the sand- fill blow into the empty hole of a stope. Whatever kind of pump they were using I knew had to be strong, because that sand-fill was moving at a high rate of speed, and I'm sure gravity added greatly to the velocity of the mixture as it dropped down the eight-hundred-foot main shaft.

I ended up just sitting or, more accurately, lying there for two weeks, occasionally calling the surface to give a progress report. Other times the sand-fill plant operators on the surface would call me to see if I was awake, and when they found out I was, they must have reported it to Bill, because I never saw him once during the entire time.

With the sand-fill job ended, I wondered what was next, hoping it wouldn't be nearly as boring. I needn't have worried.

A Glimpse of the Future

There was one much older miner at Section 35 by the name of Calvin Cargill. I didn't know exactly how old he was, but it was a lot older than anyone I expected to see working there.

On the Monday following the completion of the sand-fill job, Bill assigned me to work as a helper for Cargill. I didn't know it, but Cargill's helper had been promoted to miner, leaving Cal without a helper. It was a great opportunity, but I knew nothing about that at the time, as I thought I would be working with Cargill for the day only.

Cargill was working in the 502 stope, and by the time I had gotten down to the lunchroom, he was already gone. Bill told me to hustle back there, find Cargill, and do whatever he told me to do.

The 502 was only a few hundred yards from the station and thirty feet above the main drift, so I was back there and up to the stope in a hurry, where I found Cal, waiting, wondering what had taken me so long.

Oh crap, I thought, that's all I need is to piss off this guy with a hundred years of experience.

But he was pretty good about it. "Get a double jack and break up all that ore on the grizzly," he said. "I'm going up to the face to fix the slusher." With that he took off. Not really much of an introduction.

A grizzly was a large grating made out of railroad track cut into roughly six-foot sections and welded together, looking much like a tic-tac-toe game. It was placed over an ore chute opening to keep the largest pieces of ore and clumsy laborers from jamming the chute mechanism below. That didn't always work, as I already knew.

Normally to break up the pieces too large to get through the grizzly, the helper used a double jack, but if the pieces were larger than a double jack could handle, a chipper was used. I was using a double jack for this job, and I didn't know how to use a chipper anyway, had there been one around.

I got to work on the big stuff lying on the grizzly. I found swinging the double jack to be a little tricky, as it required balancing myself while standing on the rails and pounding on the rocks without falling. Slipping off a grizzly rail while hammering ore wouldn't have sent me down the chute because the openings between the rails weren't wide enough, but it sure would have hurt.

After successfully breaking up the ore on the grizzly, I started on the surrounding ore, sending everything down the chute. When there was nothing left to pound on, I went looking for Cargill.

Cal was up at the face drilling a hole for a rock bolt to which he would then attach a block, through which ran a heavy cable. This was connected to a slusher bucket used to pull all the recently blasted muck to the grizzly.

"Get all that rock down the chute?" he asked.

"Yup."

"What about all the ore around the griz?"

"Yup, got that too."

"All right, I'm Cal. I know who you are," he said, and that was it by way of introductions—a "show me you can do some work first" kind of guy, I supposed.

For the rest of the day, I followed him around doing whatever he told me to do. Mostly, that was fetching pieces of equipment and lifting timber and so on. All in all, a good day and, I liked Cal to boot.

I enjoyed my day helping Cal but assumed his regular helper was just off for a day and I would be on to other things. What I didn't know was that Cal's helper had been promoted to miner, so I ended up working with Cal the rest of the week. Much like the first day, my remaining time with Cal was enjoyable. He did the thinking, and I did the lifting, for the most part.

That last Friday after work, I was in the Dry changing when Cal approached me and said, "You're with me now; see you Monday."

Shocked as I was, I could only reply, "OK. See you Monday."

Now that was a great weekend, knowing the whole time I was being promoted to helper. Frankly I didn't know a miner could promote

Underground and Radioactive

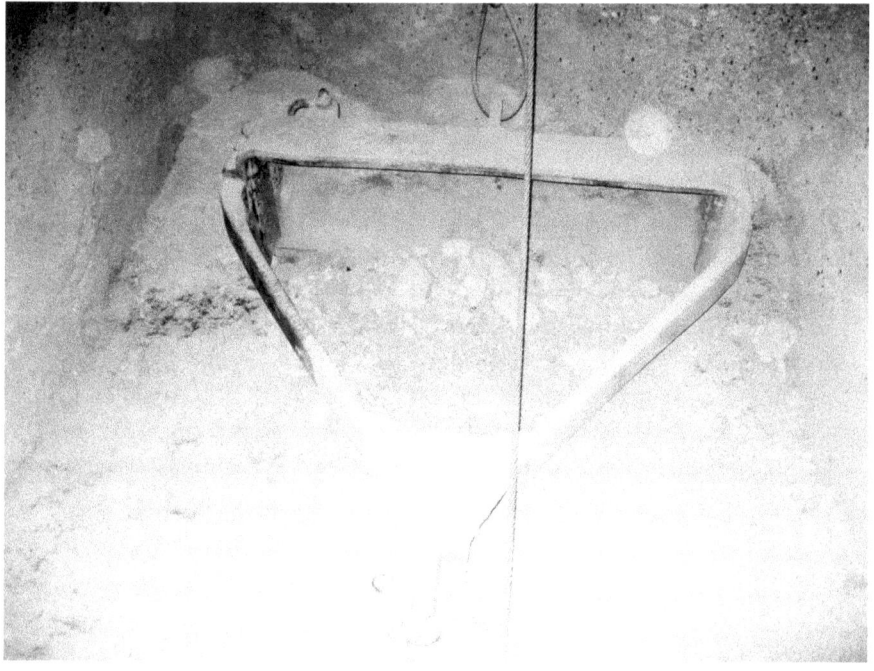

Slusher bucket. A very simple and effective device, it was used to drag ore from the face to the grizzly and down the chute. This was the exclusive method used to remove recently blasted ore from every stope (photograph by R.D. Saunders from an exhibit courtesy New Mexico Mining Museum).

someone from laborer to helper just like that. In reality he couldn't, but it sure was a nice weekend.

I was still pretty happy as I drove out to the mine on Monday and was kind of floating around the Dry when Bill came up to me and said, "I know Cal told you that you're his new helper, but you're not."

"Why not?"

"Shotgun doesn't want you helping Cal, so see me in the lunchroom and I'll come up with something for you to do."

That sucked. If Shotgun didn't think much of me I would never get anywhere. And here I thought I'd been doing a decent job of it the past few weeks.

I was finishing up changing when Cal approached me and asked if I was ready to go to work. "Bill just told me I'm not helping you," I said.

A Glimpse of the Future

"Yeah, you're my helper."

I was happy again as I went out the door toward the cage.

As I was waiting for my ride down, I noticed Cal wasn't around. When I got to the lunchroom, Cal still wasn't there, so I just waited. Seeing me, Bill came over and told me to just hang out in the lunchroom awhile. He didn't explain what was happening. I sure didn't know, and I don't think Bill did either.

Meanwhile, up on the surface, a discussion was taking place between Shotgun and Cal concerning who would be helping Cal.

According to Shotgun, it wouldn't be me, but according to Cal it would be, and furthermore, if it wasn't, then he, Cal, wouldn't be working until I was his helper.

Believe me, I don't know what I could have done to deserve the special treatment, either from Cal or from Shotgun, for that matter. There were plenty of laborers with more time on the job, but for whatever reason Cal wanted me, though I'd only worked with him for that one week. It could be that my relative lack of seniority was the reason Shotgun wanted someone else to help Cal.

Shotgun didn't hold out long. Cal simply refused to budge, and Shotgun, not wanting to lose a hand like Cal, eventually consented, making me Cal Cargill's new miner's helper.

Shortly afterward Cal showed up in the lunchroom and said to me, "Let's go," and that was that. It was the start of the best period of my mining career.

Encino

I came to learn that my college roommate Gary Mitchell had returned to New Mexico for a short stay with his family before starting law school in the Seattle area.

He came from the small town of Encino, New Mexico, about two hours southeast of Grants.

Wanting my visit to be a surprise, I called his home and spoke to his mother, letting her know what I was planning. She thought it would be a good idea and that Gary would be happy to see me. This was over a weekend, so after work on Friday, I left immediately, hoping to reach Encino before nightfall.

Encino is a place on the eastern plains of New Mexico that I had heard much about but had never seen. Much of the surrounding area is grassy and good for cattle ranching. Few people lived there at the time, and few live there now.

There wasn't much in Encino other than a highway department facility, post office, schools, and a gas station. During the time Gary was growing up in Encino, there had been a very small high school. In fact, he had as many brothers, four, as he had male classmates. Even so I managed to get lost looking for the place.

Although a ranch was large in acreage, its entrance could be difficult to locate, I discovered, so I had to stop at the gas station and ask directions to the Mitchell place.

I found the Mitchell ranch easily after that and drove up to see some men, including Gary, working around the barn area.

My surprise worked very well, as he didn't even know I was in New Mexico, let alone planning to visit his home. I was introduced to his brothers, although I believe one of them was elsewhere. Both of his parents were very friendly, welcoming, and accommodating. Sleeping on the floor would have been good enough for me, but I got my own bunk. But first I had a little lesson to learn about life on a ranch.

Just as I was thinking, When do we eat? someone said, "If you want to eat, you have to work." For about a second, I thought that was very funny before I realized whoever said it had been dead serious. On that ranch you would work first, then eat. The first job was picking up a load of hay for the horses.

The hay they were buying was a set price for a pickup truckload full of hay, and in those parts when someone said a pickup load, let me tell you that meant every square inch of space on the truck. If it wasn't already occupied by the driver and two passengers, then there was going to be hay in that space. And a load meant far more weight than the truck was rated to safely carry.

With Gary driving, one of his brothers and I got into the truck and together we drove to the sellers' place, arriving shortly before sundown. That part of New Mexico out on the plains can get mighty cold and windy during the winter, as it was that late afternoon. I wasn't expecting to work that weekend, so I wasn't dressed for being outside in freezing temperatures and icy winds.

As we started loading bales into the back of the truck, I warmed up fast, though, and we didn't stop loading until the back of the truck was just a few inches off the ground.

When the three of us got back into the cab for the ride back to the ranch, the view of the road was much different than it had been prior to leaving the Mitchell ranch, as the front of the truck had a definite upward angle to it.

I hadn't noticed the ice and snow on the road before, but now, with the truck overloaded and already wobbly, I began to worry about what would happen when we hit a patch of ice.

Gary had obviously done this hay run before and knew all about how to drive a severely overloaded truck. Taking no chances, he drove us back to the ranch very, very slowly.

When Gary parked the truck by the barn, I expected to start unloading it. By this time, it was after seven o'clock, and I was certainly hungry but still expected to have to unload the hay into the barn before dinner. Mercifully that didn't happen, and we all went into the house for a great dinner that his mother had prepared while we were gone.

Underground and Radioactive

I have no memory of what she made, but I do recall how much I enjoyed it.

After dinner everyone sat around telling stories, and I learned more about life on the ranch. I slept well that night.

The next morning Gary, two of his brothers, and I began unloading the hay from the pickup. This involved attaching large hooks on the end of a cable to the bales and hauling them up to the second story of the barn by way of a small hoist.

It was sunny that day but very cold. I got the job of standing in the back of the truck and attaching the bales to be winched up to the barn. I assumed that was the low-man-on-the-totem-pole job because it was freezing standing in the back of the truck, exposed as it was to the cold and wind.

After finishing that job, we had to do a few more chores around the barn, and then it was time for lunch. While we were eating, the brother who was either studying to become a mechanic or already was a mechanic volunteered to give my El Camino a tune-up.

My car seemed to be running well, but I was assured that it would be better than ever, and he would be doing it all for free. It's hard to pass up a deal like that from a mechanic.

After the tune-up was completed, it was around dusk, and I needed to be getting back to Grants and ready for work the following day, so, with many thanks and good-byes, I started back.

Gary Mitchell went on to a career as an excellent attorney for many years in New Mexico, and as of this writing, still is. Whichever brother it was who tuned up my vehicle, on the other hand.... I sure do hope he studied hard at mechanic's school and that things turned out well for him, but that El Camino ran so poorly all the way back to Grants that I was lucky to have made it.

The next morning the car wouldn't start, so not only did I have to have it towed to a mechanic, but I was of course late to work.

When I finally did arrive, my new partner, Cal, was not too happy with my being so late and, as we labored, proceeded to spend about an hour telling me all about how much work we had to do.

Cal

There are people who pass through our lives, however briefly, who change us forever. They can be those we have known for years—parents, husbands, wives, children, siblings, or friends. They can be someone we don't know or someone we know only briefly. Their impact is profound and enduring, emotionally, spiritually, physically, and sometimes more. They come into our lives and then, we being powerless to prevent it, they go. For me, Cal Cargill is one of those people.

A very high percentage of the actual ore extraction we did was primarily physical exertion with minimal mechanization. There were no automatic drilling machines, no large rubber-wheeled mucking vehicles, no continuous mining machines. We drilled and blasted, the rock was broken by hand, and the muck stick (shovel) and double jack were constant companions.

That kind of work takes a toll that adds up over the years. At an age when most miners had moved on to supervision, if they were still working underground at all, Cal Cargill carried on at age fifty-seven. He was by far the oldest miner at Section 35, and he looked it. At first sight I had guessed Cal was at least fifteen years older than anyone else at Section 35.

He was not particularly physically imposing but at a wiry five foot ten or so had a look about him that I found sobering. He was tough and all business with a sinewy exterior and generally hard facial expressions that, taken together, probably explained why most of the men kept their distance.

Years underground had made him the kind of leathery, grizzled-looking miner I'd heard about, read about, and imagined but certainly had never seen. It was not long before I found looks were the least of it. His countenance was a recurring reminder of how little I knew.

Underground and Radioactive

He had spent thirty-seven years alternately and at times concurrently logging and mining gold, silver, and copper around Butte, Montana, so it was extraordinary that Cal had avoided serious injury and indeed remained relatively unscathed.

Underground uranium mining was in many ways rewarding. It was always a good feeling coming to the surface knowing I had put in an honest day of hard labor and was getting paid for what I accomplished. At the same time, it was very satisfying being around so many others who were proud of what they did. Mining in the manner we were was grueling, perilous work, the physical demands of which made it somewhat rare to see men in their forties, let alone fifties, still on the job. Yet somehow Cal had done it and, as I discovered soon enough, not altogether by chance.

Though he had survived a long career in mining, by the time he showed up at Ambrosia Lake, Cal no longer had much interest in extending himself through constant physical labor, an awareness he remedied by taking on partners who would and could. It wasn't that Cal couldn't do the work, because, as I saw on many occasions, he could. He just chose not to on a daily basis.

While many of the more experienced miners were in effect foremen of underground teams, Cal was too contentious, too much the individualist, and too stubborn to put up with the many personalities on a crew. Instead he commonly took on a single partner, and the two of them worked alone.

To work with Cal for any length of time as a helper required three principal necessities: loyalty, physical strength and the ability to keep your mouth shut. There might have been other criteria, but whatever they were never became apparent to me. He certainly didn't demand any prior knowledge of drilling, blasting, or mucking. I was living proof of that.

He never required of me any thinking or problem solving of any kind. Cal let it be known that thinking was his job. He didn't need a conversationalist either. The less said the better, and whatever I did say had better have a point. No jokes, no stories, no idle chatter. Essentially, I should keep quiet, and all would be well. No problem for me there.

In the early days of my time underground, many of the miners I came into contact with assumed I didn't know much about mining or manual labor. That was a fact I didn't try to hide, and when I did, I regretted it anyway, as in the earlier stuck chute caper.

During my short underground career, I never bragged about things I hadn't done or mentioned my inexperience but at the same time infrequently admitted what I actually had done, many of which were embarrassing episodes. I was a greenhorn hand, and although I tried not to be, this was something I thought about often and found unsettling to be reminded of, so I kept my mouth shut. I didn't know it yet, but the older hands in some ways respected this in a new guy. This was especially true of Cal.

After I came to work with and to know Cal somewhat, he would remind me almost daily that no matter how long I worked there, I would never know more than he'd forgotten. Yet he assured me that one day I would be a good miner since all I had to do was try, "Because that's all a mule can do," he would tell me. That sure sounded more like me all right. But that was some ways off, and as a lowly laborer observing Cal from a distance, he initially struck me as someone I didn't want to be too close to for any reason.

As laborers we were occasionally, though rarely, allowed to work alone, but miners and helpers were never allowed to work without each other. You had to have a partner with you at all times as a safety measure. If your partner called in sick, you either couldn't work or were assigned a temporary helper for the day. Most miners' helpers very seldom called in sick for fear of losing their place. Sometimes a helper was promoted to miner after a period of apprenticeship, as had been the case with Cal's helper and the source of my big break.

What made a helper happy was how much contract time the miner he worked for gave out, contract time being the hourly rate a miner made based on how much work the pair did and how much ore they produced. If a miner had a set hourly rate of, say, $20.00 an hour, then what he accomplished each day was added to that, the total of which determined his contract rate. Kermac rules allowed a miner to designate up to four hours of contract rate per day to a miner's helper.

While the helper contributed significantly to production in a stope, some miners were stingy with their contract time and only gave an hour or two per day to the helper. Cal, on the other hand, would

Underground and Radioactive

While it could just as well be from Section 35 in 1976, this excellent photograph from 1909 illustrates what underground square-set construction looks like. It is an extremely effective means of ground support. Invented in 1860, square-sets were first used during the Comstock Lode silver boom in Nevada (John Oxley Library, State Library of Queensland, Australia. Author unknown).

always give the maximum four hours to his helpers whether they were worth it or not, and if they weren't worth it, they didn't last long around Cal anyway.

Although I had been with him only a week, Cal had been drilling, blasting, and building square-sets in the 502 stope for several months. As I made my way up the short twenty-five-foot manway that first day as his official helper, I was thinking about how interesting this experience might be.

Cal, already in the stope, was waiting for me as I topped the manway with a stern look that concerned me at the time but which I later learned was nothing personal; it only meant we had a lot of work to do. This of course came to mean how much lifting and hauling I was going to be doing. His advice to me was then, and always remained, to "work like a mule and always be doing something even if it's wrong."

Cal needn't have reminded me, because I'd already gotten the message, but he went on to say something about how we had to get going and how much work we had to do. During the previous week, I'd heard a variation of that every day, but it summed up Cal. He tended to repeat himself over and over. If I repeat myself too often in this narrative, blame Cal.

During a typical day with Cal, we hauled a lot of timber and built several square-sets for ground support. I worked hard, carried a lot, and did everything Cal asked. My unintentionally being unable to answer any question or detail about real mining earned me my first "I've forgotten more about mining than you'll ever know" lecture. I took no offense to it no matter how many times he said it, and he said it a lot.

I got a free introduction that day on how timber stopes were built in 1961 and how poorly some miners at Section 35 were building them. I think the point of that was that he expected me to understand the importance of ground support and the principals of set construction. Perfection was emphasized in that everything must be plumb and level if it was expected to support the ground above. Then I got the "how bad the ground was" lecture.

Cal was very concerned about the ground we were working in, how noisy it was, why we were idiots for working in it, and how he was going to quit any day now because of it. When he asked me if I knew how bad the ground was, I said I didn't. Then in his uniquely nonjudgmental, almost humorous way, he told me about a dog he once had that wasn't that dumb. I learned a lot about that dog over the next several months. Evidently I was Cal's favorite source of underground entertainment as he always got a big kick out of how little I knew about mining.

One day I was walking along the lagging among the square-sets about three stories up when suddenly in an instant I found myself on my knees, having no idea what had happened. I hadn't felt the hit, but a slab from the back had fallen and landed squarely on my hardhat, knocking me down. It was quite a shock for me but amused Cal to no end. It was about that time that he really started to complain about the condition of the ground above and around us.

What Cal was telling me about the ground we worked in was true. Although most called Section 35 a hard-rock mine, we never actually worked in hard rock at all. It was mostly sandstone and part of the ancient prehistoric lake bed now known as Ambrosia Lake. In fact, the hardest material I ever drilled into was the occasional dinosaur bone that we would encounter, the largest of which were on display at the Kermac district office.

Sandstone breaks up fairly easily, especially when compared to granite, for example. To the miners who had worked in real hard-rock mines elsewhere, Ambrosia Lake was a comparatively dangerous place. They all knew it was risky, but in mining everyone went where the money was.

On the other hand, there were people like me who had no idea what a hard-rock mine was, so we knew no better. As to the conditions we were working in at Section 35, I considered myself at least wise enough to rely on the word of much more experienced miners.

When a large cavity is created while mining, it takes an assortment of support structures and techniques to keep sandstone from falling in. Various methods were tried, but the most successful was the square-set, and at the moment the most experienced square-set miners were coming from the old mines in Butte, Montana. That's where Cal Cargill and his knowledge came in and why he was here at Section 35.

It's also where Shotgun Buchanan came in, as he too was an expert at square-set mining. It was no accident that Calvin Cargill had come to Section 35. Cal and Shotgun were old acquaintances from having worked together in Butte for many years. I always had the feeling from the way Cal talked that he and Shotgun had been pretty good friends at one time, but it didn't seem so now. Whatever the case was, there was enough history between them that Cal was able to convince Shotgun to assign me as Cal's permanent helper. Many years later I came to learn that Cal's son was married to Shotgun's daughter.

Cal was tuned in to whatever noises the shifting sandstone was making in our stope and cognizant of the overall feel of things there. I didn't hear much of anything. He saved both of us a few times and once in a particularly dramatic fashion.

We had been working in the 502, drilling, blasting, and building square-sets until it looked like the skeleton of a new office building going up. It seemed to me about the size of a city block, with square-sets up to four stories in some places.

As we built it up, Cal continued to complain about how bad the ground was. I still had no idea what he was talking about but assumed he was right. He said many times that the stope wouldn't be there long, never sounding worried but always seeming to be stating a fact.

Listening to Cal talk like that almost daily, I started to dislike working in the 502 myself. I began having nightmares of being trapped. It took many months of working there to figure out the combination of reasons that made it a bad place to be.

Although each stope had only one manway, there were usually connecting drifts to other stopes that, although difficult to access, provided escape routes. Sometimes they were just crawlways, but at least there was another way out should the manway become blocked. When working a stope, everyone I knew was alert to where the escape routes were at all times. The 502, on the other hand, had no alternate escape route. It was the manway or no way.

The closer to the main shaft, the older the stope was, and the 502 was very close. That meant there had been mining going on in there for several years, accounting for its large size, irregular shape, and odd personality. Every miner who worked it had a different style. This was especially evident in the hodgepodge of ground support systems that had been erected over the years.

Some miners were perfectionists. In a timber stope, it meant they made perfect cuts and used carpenter's levels when building square-sets, while other miners might throw anything up as fast as possible and move on. To them it was all about filling the ore chute and not much else. They made great money, and some survived unscathed this way, but some probably didn't. The best timber-set miners built close-to-perfect square-sets.

There were miners who worked far in front of their ground support, leaving large open spaces. They'd use stulls or roof jacks as temporary support for far too long. That was a dangerous practice and exposed them to falling slabs of rock. While stulls did work extremely well in smaller, relatively confined spaces, they had a specific purpose that did not include large open areas.

The stulls we used were wooden posts about the circumference of a telephone pole and on average from five to eight feet tall. We would cut them to size to support the ground directly above where we were working. Ideally that meant right over our heads. For a stull to work as intended, it, like the posts of a square-set, must be precisely vertical.

Imagine lifting a car with a jack that is at as little as a ten-degree angle. The car is going to fall off the jack, and hopefully nobody is underneath when it does. A carelessly installed stull leaning at ten degrees has about the same chance of holding the ground above as the jack does of holding up the car.

A roof jack looks like a very tall automobile jack. We would put a four-by-four piece of lagging on top of one, then jack it up to the back as tightly as possible. They were meant to be temporary solutions to back support, but some miners used them far too frequently as permanent support. As always they did it to save time and produce more ore, but it was dangerous as a permanent back support solution and sometimes served to destabilize the surrounding area. I saw many bent and bowed roof jacks that had been left in place long enough for the ground above to destroy them. I can only imagine how much Kermac spent on misused roof jacks.

The 502 had a combination of various types of ground support and was a reflection of the many miners who had worked in it. I could tell where one very good square-set miner had started building and another very poor square-set miner began building.

In other places previous miners had forgone square-set construction altogether, having instead thrown up a forest of uneven stulls. Those stulls were twisted into many odd shapes by the ground above slowly crushing them. Coming upon a large collection of stulls being contorted in that manner was an eerie sight.

From what I had observed, stulls and roof jacks were both methods of ground support most likely to suffer sudden failure if used over the long term.

Timber stopes were noisy. Even I could hear it as the weight of the earth compressing the wood in square-sets and stulls creaked and popped fairly often. It was somewhat unsettling, but I got used to it, more or less.

As long as a stope had ore, it was going to be worked until emptied of ore and eventually filled with sand. In an older stope like the 502 with a tremendous amount of ore to be mined, that resulted in a lot of timber being used. Some of it was very old, and it became evident what the pressure of the earth above could do over time. In places the weight had crushed twelve-by-twelve-by-twelve timber to just a few inches high. Stulls in many places had either been compressed a foot or more or snapped in two.

While eventually square-sets were crushed from the weight of the ground above, for the most part well-set timber did a good job of ground support. The sets were extremely strong and, when built with care, kept the back from falling in incredibly well—at least enough as to allow the extraction of the ore body. On rare occasions timber sets did fail suddenly and catastrophically.

As we were getting ready to head off to lunch one afternoon, Cal said to me, "Let's get out early today; this place won't be here when we get back."

"What?"

"Let's go in early. The whole place is coming down."

"The entire stope?"

"Yeah; let's get out of here early."

So we left for the lunchroom and a nice leisurely meal, during which I didn't ask any questions but wasn't convinced that Cal could predict the collapse of an entire stope. Sure, he knew a lot, but this?

An hour later we slowly walked back to the 502. As we approached the manway, something didn't seem right. Actually, something didn't smell right. There was a very earthy and damp smell in the air.

The manway was clear, so Cal made his way up first as usual with me following close behind. As he reached the top, he looked down and said, "You have to see this, pard."

The 502 was a mass of rubble, boulders, and timber. Hundreds of square-sets were down. It was an incredible sight.

Cal had been right: the stope was gone. Up until then I'd been faithful to the rule of seldom asking questions, but this seemed especially important and worthy of an exception, so I asked Cal how he knew.

He said he'd heard it collapsing before we'd left for the lunchroom. I'd heard nothing or, more likely, recognized nothing I heard. It was what thirty-seven years of experience working underground meant that had saved us. It scared me then, and I often think about what could have been had I not had the good fortune to have been working with someone like Cal.

Cal's reaction was more matter of fact than nonchalant, but he definitely didn't seem the least bit shaken. I don't think he was fearless; he just knew what he was doing and that things like this happened underground. Following Cal's lead, I tried my best to act as if it was just another day, but I was feeling neither nonchalant nor fearless.

Later, after I became a fully certified miner, I too had parts of stopes collapse, but nothing that large or even close to it. I never heard it coming, and fortunately it always happened when I wasn't there. Instances like this further shook me in that I'd still not recognized whatever warning signs were crackling in the air around me. I had just lucked out.

These events left me fearful for a few days at a time. They further solidified my acknowledgment that I really didn't know what I doing and felt as if I never would. Trying like a mule and doing things even if they were wrong, combined with simple desire, was not going to suffice. It seemed my best prospects for survival underground had been reduced to luck and Cal. I didn't have much mining talent and even less intuition. Of the warning signs all around me, I was unaware of most and evidently ignored the rest.

It was 1976. I was an immortal twenty-four-year-old making thirty dollars an hour with no bills, no responsibilities, and few cares. What had transpired to date wasn't going to scare me out of something I really enjoyed doing. I still had the camaraderie that those of manual labor often share, was thrilled to be going underground every day, and was proud of what I did.

The 502 having collapsed, Cal and I were moved to the 803 stope, where we started all over again setting things up the way Cal wanted them. I did the setting up really, and it was Cal who did the directing, which again was OK by me.

I was determined to learn as much from Cal as I could. It was not easy. His answers to my few questions were usually short yet always to the point. Sometimes he wouldn't answer but would quietly show me, maybe at the time I asked, maybe later. Often I was frustrated learning primarily through observation. If I was not learning what I needed to become a decent miner, I was learning to listen and observe.

The price of knowledge was a lot of heavy lifting. Cal, being not as interested in doing nearly as much heavy work as was needed, had

Cal

me to do it. I carried all the supplies and set everything up in the stope that needed to be done. Cal did the thinking and the planning. If I ever had an idea and was brave enough to bring it up, he never said it was a bad idea; rather he always let me know it wasn't a good idea for me to be thinking and that it was he who had the experience.

Cal expected his helpers to watch his back, and he would watch theirs. This meant both on the job and off. Miners did a lot of roughhousing, and some of it was pretty intense. One incident in particular stands out.

Thirty or so miners were standing around the headframe waiting for the cage. One of the younger, hotheaded miners, Larry "Boots" Randolph, loved starting fights for no apparent reason. The odd thing about it was he never did it in anger. To Boots it was just a game.

Boots couldn't have been more than twenty-five years old. I knew him slightly, having carpooled with him for a short while until I found that he took it as a personal insult if anyone had the audacity to pass his 340 Duster on the drive to or from work. We got to work fast, but it wasn't worth the risk, which I deemed far in excess of anything I experienced underground. The only question was when he going to wreck, and I didn't want to be around when that happened.

For some reason as we were all standing out by the cage, Boots got it into his head to start something with Cal. He jumped on Cal's back and started what he thought would be a one-way wrestling match. Old Cal might not have appeared as if he could do all the heavy lifting eight hours a day, but he was one strong and nasty old coot who had been in enough fights to equal ten times Boots' age. In this case he'd forgotten more fights he'd been in than Boots would ever have. It was a severe miscalculation by Boots to have expected Cal to appreciate any roughhousing, even if good-natured.

Cal turned on Boots so fast that soon young Boots found himself on the ground being administered quite a beating, no doubt wondering what had happened. It took a couple of miners standing close by to break the whole thing up.

I was standing some ways off watching and was taken by surprise myself. Astonished that Boots would try such a thing, I saw how fast Cal moved and found it amusing to see Boots sprawled on the ground, having been taken down by a man more than twice his age within a few seconds.

Underground and Radioactive

It seemed to be all over in thirty seconds, after which the cage arrived and we all went down. I thought that was the end of it. Boots wasn't the only one who had miscalculated. It might have been the end of it between Cal and Boots, but there was no end in sight for me.

While I found the incident immensely entertaining, Cal did not and spent the better part of a week berating me for not coming to his aid. Someone had jumped his back and all I had done as his partner was stand there. In fact, he got so angry I thought I might be back to laborer and under the wing of Frankie or Bill again real soon.

Day after day he asked me why I hadn't done anything, to which I didn't had a suitable answer until one day it came to me.

Cal was again in the middle of a tirade executing a lesson plan on how a partner always protects a partner's back when I stopped him and said, "Cal, if I thought you'd needed any help, I'd have given it, but you didn't need any help with that idiot Boots."

And that was all it took. Cal hesitated a moment after hearing that, then seemed pleased, and that was the end of the tongue lashings. That incident had cost me several days of misery, and now I was going to make a point of talking to Boots about it the first chance I got.

Now that Cal and I had things squared away, I confronted Boots in the Dry that same afternoon, telling him never to mess with Cal again or me either, for that matter. With that said I whirled around with an extended elbow and clipped Boots hard in the nose.

I just didn't do things like that, so it surprised me, let alone Boots. The elbow didn't do any damage, but Boots wisely let it go, and we never had a problem between us again, and he never came close to Cal.

Cal was fond of telling me about life in Montana. He wasn't only a miner but also a logger. He loved talking about logging in the cold. When he said it was sixty below, there was never any talk of wind chill. It was just plain sixty below. He talked about diesel fuel and oil freezing and trees exploding in the extreme cold.

He talked about being buried in snow while in his home and digging out and then going to dig neighbors out. Then he talked about the heat in the summer. I'd been to Montana in the summer and knew about the heat. You'd never expect it to be that way, but it could get blistering hot.

I gathered that Butte was a rough place to work and live, as Cal seemed to have had quite a background in fighting. Yet he was actually not aggressive at all and one of the least crude-talking miners I ever knew. It was never his habit to curse anyone or anything as a starting point.

He seemed to be something of a firearms expert, often hinting in the vaguest possible way that he might have scared off an unpleasant former associate or hitchhiker by showing a pistol. I was never sure; I didn't want to know anyway but never doubted it. I just never had any reason to doubt anything he said.

Cal mentioned that during his mining career, he had been caved in on, buried, and caught in blasts, always adding ominously that through it all he had still never seen ground as bad as we were now working in. His stories added color to his formidable aura.

Cal was a unique character who could look at you in a way that adjusted your behavior so that whatever it was that had prompted the look would never happen again. If only college, or "book learning," as Cal said, had been so uncomplicated, I would have graduated in a year.

In my previous jobs, I had been taught to always make it appear I was working, but now I was learning something new. I could never stand around watching, seeming to be busy or appearing to be ready to work. If I did, Cal emphatically explained what I should be doing.

Once, imagining there was nothing to do, I mistakenly asked what I should be doing. That was a blunder, as he let me know I should always know what to do next.

Letting Cal know I was a college graduate was a big mistake too, for this only made me a smart college boy working in a mine learning to be stupid. Cal then went on about how a smart college boy should be able to find something better to do and to get to finding out what it was. Well, I thought I was, but I certainly wouldn't have tried explaining that to Cal.

Many times during one of Cal's "you always have to be doing something" lectures, I was inwardly smiling and thinking about a Gomer Pyle episode. During boot camp Sergeant Carter often had Pyle dig a hole, then fill it up and start over. This amounts to doing something while doing nothing and is a valuable skill no matter what line of work you are in. It might be wrong, but at least you are doing something. I frequently thought of my and Cal's relationship in those terms.

Underground and Radioactive

Eventually each day I learned what we would be doing the following day and anticipated what it was that Cal would be needing. I got so good at it that I had tools and supplies ready before he asked for them.

When I saw we were almost ready to drill, I would set up the drill, for example. I surprised Cal a few times this way when he said we were going to drill and I replied everything was all set and ready for him to begin. In this way every now and then I'd get an approving look. Those looks were better than a pay raise and made my day. I could see why people who worked for Cal loved the guy and became good miners.

Although we had geologists in the mine searching for the best ore deposits, Cal "pictured the earth," as he liked to say, and just as a good fisherman knows where the fish are, he would find the uranium ore. Young "college boy" geologists visited our stope and, using probes, pointed out where the good ore was and in which direction we should go. Cal's indiscriminate response showed he held only slightly less contempt for geologists than shift bosses. I liked the geologists, but around Cal I just kept my mouth shut.

Usually the geologists were right, but Cal often went the other way and was always right too. It isn't easy to do this. Raw uranium doesn't glow and doesn't usually look like anything other than plain old average rock. But Cal knew what it looked like or smelled like, or maybe he had a mental Geiger counter. Whatever it was he had, I didn't have it, and neither did most other miners I knew.

After we had gone our own way, found rich ore, and made a lot of money getting it out, I would periodically ask Cal how he knew. He might have answered a time or two, but I'd usually get a look and spend the rest of the shift comparing where we had mined and where the geologists said we should have mined trying to figure it out.

"Try—that's all a mule can do," Cal repeated over and over. It took me a while to understand this bit of wisdom, but I think it came down to a combination of perseverance and always doing something—even if it was wrong.

Obviously Cal had spent some time around mules. I took it mules understood certain commands, usually short one- or two-word commands. It didn't take me long to draw the comparison between the commands a mule would get and those I was getting.

Cal

College education works. I quickly caught on to the meaning of one- or two-word commands. For my own peace of mind, I decided mules couldn't be dumb animals. In fact, they must be very smart, highly developed mammals, just like dogs. They don't think much, though. They just try.

With Cal, if what he tried didn't work, it was on to something else. Never wasting time fooling around with something that hadn't worked, he just kept trying things until something did.

Ultimately the day came that I was promoted to miner and was given a certificate to prove it. It could only have been at Cal's urging.

Cal had decided to return to Butte and semi-retire. He said that seeing as the ground we worked in was too soft and too dangerous, he preferred to get back to his dogs and mules and maybe do some logging.

In retrospect perhaps the prudent thing to do would have been to follow his example and retire myself, but I truly loved being underground and working the way we did, so that thought never occurred to me. No longer having access to Cal's knowledge and advice bothered me, but I recognized that things were always changing, and people come and go.

Although I had been with Cal for almost a year by then, he had to have known I was not truly ready to mine on my own. Certainly I wasn't convinced of it. I think he must have put in a good word with Shotgun and Mel Vigil, though, so they went ahead and promoted me.

In his many years underground, Cal had made few mistakes, and recommending me for promotion to miner might have been one of them. I was not ready.

I was Cal Cargill's final mining partner. He did retire from mining and never again worked underground. Cal must have worked with hundreds of partners in his many years of mining, and I was proud to have been one of them and honored to have been his last.

Although I wasn't prepared to mine on my own, turning down a promotion to miner was out of the question. I was recommended by Cal Cargill, the roughest, toughest, meanest miner and mucker. Who could turn that down? Was it possible I knew more about mining than I thought I did? Probably not. Still, I could try, work like a mule, learn on the job, and do whatever it took, even if it was wrong.

Underground and Radioactive

Cal and I had our only meal together at the Frontier Steak House in Milan the evening of the last day we worked together. The following day he left Grants, returning to his home in Butte.

For me there was nothing bittersweet about it. Without Cal I was heading for trouble and felt lonely and miserable that night. I hope he never sensed it, but if he didn't know what a greenhorn college boy like me was thinking, it would have been the first time.

He spoke a lot that night about his love of Montana and being anxious to get back, of how his health wasn't very good anymore, and again of how bad the ground was where we had been working and how I needed to watch out for it.

There were a lot of technical questions about mining that I wanted to ask but did not. If I hadn't asked all those questions in the stope with the answers right in front of us, what good would it have done to ask sitting there in a restaurant? Maybe anxiousness was all it was holding me back, or was I more skilled than I thought? Certainly all my questions would have only served to weigh on his conscience anyway.

The meal over, we shook hands in the parking lot, and then he was gone and I was alone. I never saw Cal again. I'd learned a great deal from him, but too little of it had to do with mining.

Alone

Mel Vigil was the one who actually handed me my promotion papers, and when he did so, I had the feeling he thought a little more of me than he had the first time I had stood in his office. Oh, that bemused look was still there; it just was not as obvious.

My loneliness, insecurity, and fear were temporarily displaced by hopes and dreams of what for me would be colossal paydays.

Everyone at Section 35 made a point of checking the contract rates that were posted every other Friday in the Dry. Posting those rates for everyone to see was a further incentive for each miner to work harder, earn more, and look better. Posting rates might even have contributed to so many miners ignoring time-consuming safety regulations. No miner I knew ever wanted to be on the bottom of that list, including me.

The top contract guys were doing things nobody else could do at Section 35. They were the elite. Some exploratory work and raise and shaft construction paid upward of $300 an hour.

Other than greenhorn laborers, they were the only ones in the mine who did the job nobody else wanted. They volunteered for it, knew what they were doing, and were paid well for doing it. They took enormous risks working in unsupported areas, drilling out access shafts, air vent shafts, track drifts, and main shafts.

They frequently and by choice worked in areas with no ground support. Working very quickly, they produced so much that no shift boss dared to mention safety rules or violations. These were the miners who were more or less independents. The company paid them, but they worked by their own rules. Everyone at the mine held both the miners and the helpers on these crews in the highest regard.

The elite crews at Section 35 were always made up of one or two miners and several helpers. It was dangerous work for some helpers, being on one of these crews, but they made more than many miners did.

Seeing where they worked and how dangerously they did it, I wanted no part of one of those crews. I would have only endangered them and myself. And, oddly enough, I never saw a single fatality or injury come from one of the best crews.

The really proficient pillar stope and timber stope miners averaged between $80 to $200 an hour. Below that were some fairly decent miners who might have been stuck in a poor area of the mine that did not produce as much as the geologists expected. Further down the list were the average guys at around $30 to $60 an hour, who did all right but were new, worked more slowly, followed a lot of safety regulations, or just weren't extremely motivated.

A miner who fell below $20 an hour probably just wasn't very good. I had high hopes of being somewhere in the middle of the contract rates. I didn't want to be the guy at the tail end about whom everyone asked, "What the hell is wrong with this guy?"

So it was with middling expectations, great enthusiasm, and many questions that I reported for my first day as a genuine certified miner.

What stope was open, and which one would I get? Who would be my helper, and who would be working the opposite shift from me? I was somewhat apprehensive, yes, but ready to get to work, produce chutes full of ore, and throw up the best square-sets at Section 35.

Bill Clark was still my shift boss, so when I got to the lunchroom on my first day as a miner, I reported to him, and all my questions were soon answered.

I was being assigned to perhaps the worst stope imaginable. Correction: former stope—my old friend, the collapsed 502, from which Cal and I had escaped.

Why management was still calling a collapsed ruin a stope I didn't know. Perhaps this was punishment of some kind, or perhaps it was someone's way of telling me Cal wasn't around to look after me any longer—a thoughtful reminder I could well have done without. The insecurity and fear returned.

Bill told me that it had been determined something had to be done about the 502 stope. I was to go back there, inspect it, and report to Bill for further instructions and assignment of my new partner.

Alone

I was thinking, Check what out? It's gone, Bill. Report back? Report what back? It's gone.

Inwardly I was beside myself, but I did as I was instructed. No sense in jeopardizing my new status on the very first day. Whatever I was feeling paled in comparison to what transpired concerning the 502 stope over the next several months, though.

To this day I don't know what I was supposed to be looking at. I had been there and seen it at the time it had collapsed and didn't need to see or know anything further. I couldn't comprehend anyone ever working in there again.

When the 502 had collapsed, it seemed mine management agreed with Cal's assessment of the stope being unworkable, and, despite a large amount of high-grade ore in the area, had abandoned it. I assumed by now it had been backfilled with sand and that was the end of it. It was not.

I trudged down the main track drift on my way to revisit the 502. It was a bit melancholy returning again to the place where I'd spent so many hours working with Cal, where I'd learned something about life and so little about mining.

Climbing the manway, I hoped the entrance would be caved in and blocked by now, but miraculously it wasn't. I took the miner's lamp off my hardhat and, flashing it around, peered into the moonscape remains of the stope and knew whatever the bosses had in mind for this thing had nothing to do with mining.

Little had changed except possibly even more of the back had caved in. There were still hundreds of huge timbers crushed and strewn about the place. The remains of the stope seemed to be the size of a football field. Slabs of every size and shape filled the area. There were remnants of our equipment scattered and visible here and there. I was impressed to have seen the disaster the first time and could have done without a second look.

Making my way down the manway and back to the lunchroom, I was beginning to think things were going to be even worse than I had imagined. Bill provided confirmation of that when he said I would be in charge of backfilling the 502 with sand.

Timber stopes, pillar stopes, and ballroom stopes produce ore, the production of which pays a miner, but there are no collapsed stopes that produce anything or pay anything worthwhile.

Underground and Radioactive

In the 502 there would be no drilling, no blasting, no mucking. It appeared I was not going to be a miner after all, but rather a sand-filler. I had done that already as a laborer. Now a miner was being put in charge of filling the 502 with sand. I was very disappointed and had to figure someone had it out for me. I didn't know who but wasn't about to go around complaining and asking a lot of questions. Talk about starting at the bottom. Those Friday contract postings weren't going to look very good for a certain miner. Perhaps this assignment was part of some grand design?

As events eventually transpired, putting me in charge of sand-fill again might well have been by design, because this was not the last time I was going to see the 502, and nobody knew it better.

Bill assigned Art Martinez as my first miner's helper—another kid recently graduated from high school, ready for a bright future underground as assistant sand-filler. The good news for Art was that he was being made an official miner's helper, somehow having skipped laborer. I thought they must really need bodies even more than when I was hired on. But I did wonder who was going to be unloading supplies at the station if Art was with me. Perhaps there was a job title even lower than laborer.

Although this was good news for Art, the bad news was we weren't going to be doing any mining. More good news for Art was that he didn't know any better.

The really bad news was there would be no contract hours on the sand-fill job. I would be making around fifteen dollars an hour as a flat rate. Had they been paying me for every ton of sand that was pumped into that stope I would have been very well off, but no such luck.

For the next couple of weeks, Art and I hung more of those twelve-foot lengths of eight-inch sand-fill pipe along the back of the main track drift from the station all the way back to the 502. And again it was tedious, heavy work.

With all my prior experience coming to bear, I became a true professional at connecting pipes with those special rubberized clamps. It might not seem very impressive, but the speed at which I was able to tighten the large bolts of the clamps using nothing but the ever-handy pipe wrench was, if I do say so myself, my proudest accomplishment on this assignment.

Alone

Blasting board. Each mine worker had a nametag on this board located in the lunchroom accounting for his whereabouts. When everyone was accounted for blasting could safely proceed. It appears from this photograph that I am currently out (photograph by R.D. Saunders from an exhibit courtesy New Mexico Mining Museum).

Some old mining towns today hold drilling contests and double jack contests. I wish they held pipe wrench contests, because even now I can wield that very same wrench with the same professionalism I did when I was underground.

As Art and I were hanging pipe, we ended up with lots of downtime because the availability of laborers at the station to unload supplies had become sporadic at best. As a result, many of the things we needed remained at the surface. Essential materiel used in production made it down the main shaft, but the things I needed were low on the list of priorities.

When we did get the pipe, no motorman wanted to haul it. Two guys who weren't going to be making the company, themselves, let alone the motorman, any money were way, way down the to-do list of any motorman. So, as I had done many times before, I secured a motor and hauled my own pipe, or on some occasions Art and I just walked sections of it back to the 502. It wasn't as if we had anything else to do.

Normally I kind of enjoyed walking the main track drift, but carrying something as heavy as sand-fill pipe makes it more difficult to avoid the speeding trains that wouldn't haul our pipe. If there is, or was, anyone better at dropping pipe and running for cover, it would be interesting to discuss their technique, because ours was pretty good.

When we had completed the pipe-hanging duty, the sand- fill began. As before, our function consisted of watching the hundreds of tons of sand pump into the stope.

All I really needed to secure a spot on the Section 35 wall of infamy was to flood the main track drift with thousands of tons of sand, so every now and then either Art or I would walk the drift looking for leaks, but we never found any. By this time, it was safe to declare myself pipe clamp champion of Section 35.

Sometimes we would have to adjust the pipe to shoot the sand one way or another, but otherwise we sat there chewing tobacco and telling stories, and by that time I had accumulated a few. I recounted for Art my many mining mishaps and adventures to date, which Art responded to with wide-eyed stares, seemingly amazed that I was still there to tell them. No doubt my survival skills sounded impressive to a greenhorn.

In return Art instructed me on the fine art of parking your vehicle on railroad tracks. When it came time to purchase new transportation and you found yourself a little low on cash, there was help available. "You don't buy insurance for nothing," he'd say, to which I responded, "You're kidding, right?"

"No way. First you report your car stolen, then go park it on the tracks. When a train hits it, you have yourself a new car."

I doubt Art ever carried through with a plan like that. If he had the news would have been all over the front page of the *Grants Daily Beacon*. Thankfully it never was.

I guess Art had a thing for trains, because later in his mining career, he accomplished perhaps the most astonishing feat of daring, ingenious stupidity when he stole a locomotive—a real locomotive from a real freight train that was idling on a side track. I don't know where he expected to go with it, but off he went down the track. That was Wild Turkey in action for you.

Art didn't get more than a few hundred feet before railroad employees easily regained control and Art was arrested. He missed some time at work as a result of that ill-advised stunt.

It took several weeks to fill the 502 and then finally, mercifully, it was over. With the place completely backfilled, I was able to report back to Bill that the job was done. My dues paid, I was ready for a new assignment.

Hopefully there weren't any other collapsed stopes needing a sand-fill expert. Perhaps now I could expect a stope assignment that included drilling, blasting, and mucking, and thankfully that is what happened. I was assigned to a real working timber stope, the 812, opposite Glen Hill, one of Cal's former helpers—in fact, the helper whom I replaced.

Hill was a real go-getter, a very good, zero-nonsense miner with an excellent reputation who worked as hard as anyone I knew. That was reflected by his contract rate that more often than not put him somewhere in the top half of Section 35 miners.

Section 35 was usually operated with two shifts. There was an 8:00 a.m.-to-4:00 p.m. shift, and a 4:00 p.m.-to-midnight swing shift followed. Most people worked two weeks of day shift followed by two weeks of swing shift. Days were preferred by most, but there were some miners who preferred working swing exclusively. Each stope usually had a day crew and a swing crew.

Occasionally a stope would have a single crew working it, as had been the case with the stopes that Cal and I worked. Because blasting was supposed to be done twice a day from the lunchroom blasting board, the opposite working crews, day and swing, usually attempted to time their work in such a way so the shift that was coming on had a big pile of ore waiting to be slushed into the chute. If the crews worked well together, that's what happened, and if they didn't, it led to arguments at shift change. I had seen a few of those arguments and had no intention of getting involved in any of that.

It's possible that Mel Vigil put Hill and me opposite each other because we had both been trained by Cal, but I had no way of knowing that, and it wasn't exactly thrilling news anyway. I knew enough of Glen to know he didn't share my sometimes lighthearted view of the job. Few did.

Hill was a man of few smiles and was all business. The prospect of failing to keep up with him or worse, screwing up, was nothing I looked forward to. As things stood then, knowing the relative degree of my own skill, that prospect seemed more likely than not.

While he didn't say it, I doubt whether Glen was encouraged by the good fortune of having me assigned opposite him either, as he must have had some inkling of what kind of miner I was. Not that Cal would have said a negative word to Glen or anyone else, but he (Glen) could read the contract postings as well as I, and mine were bad. Sand-fill bad.

When you were working opposite shifts in a stope, the most important thing was to never leave anything of yours for the opposite crew to have to clean up. You wanted to leave the stope set up and ready to go at the end of each shift, be it a drilling shift or a timbering shift or a mucking shift.

With Cal, if we were drilling, we would finish up, load the round, and then stack the timber for the square- set that was going up afterward. If there wasn't any or enough timber on hand, then it had to be rounded up and brought to the stope. When we returned the following day, it was all there ready to go.

If erecting square-sets, it was imperative to muck the round, complete the set, bring the ventilation forward, move the air and water hoses, and set up the drill. Doing so enabled the opposite shift to immediately get to work. If possible, having a motorman pull the ore prior to the end of a shift was a priority as well. If a motorman couldn't be found to do it, then "find a motor and pull it yourself" was the unspoken rule between opposite shifts.

In my experience good contract miners very rarely left the stope or heading a mess for the following shift to clean up. On the other hand, poor contract miners often did the opposite, and if you didn't want to find yourself in a fight in the Dry, you didn't do that.

At a minimum Hill knew I wouldn't intentionally leave the stope in disarray, so at least that much was on my side, but I'm sure he had his doubts. My job was to allay those fears, and I was going to work hard to do that.

Alone

Initially my plan was to try to stay on the timbering side of the rotation. I knew how to timber very well and what it took to efficiently collect the supplies to do it. Erecting a square-set paid well too, just not as well as ore production, but I was fine with that if it kept Hill happy.

While my first love in mining was drilling, I was nonetheless insecure in my ability to blast the rock out in the way it needed to be done so as not to leave a disfigured face for Hill or worse, unexploded powder that would have to be removed by dangerously re-drilling. That was one of the most hazardous jobs around and was usually caused by human error.

During my time working opposite Hill, I thankfully never left any unexploded powder in a round. Had I done so, Hill would have had me in there drilling it out to who knows what end. Powder was stable, but putting a rapidly undulating steel drill to TNT and a blasting cap would certainly be enough to detonate it.

A good miner can blast the face he's working on to exactly the size he needs and wants. The back and floor will be relatively flat and the ribs straight. I was amazed at how Cal could blast an almost perfectly flat floor with straight ribs and back, just about making the area perfect to stand a square-set. I still don't know how he did it so precisely.

I have often been asked if being tall made me claustrophobic underground. The easy answer has always been that if I thought an area was too small, I just made it bigger by blasting. So no, I very rarely worked in a confined space underground that made me feel uncomfortable.

As we were erecting eight-foot square-sets, Cal would usually leave a little room for cribbing but rarely much, making the standing of a set much faster by leaving less empty space to fill in. Cribbing in mining is much like stacking toy logs in a square pattern. Of course underground we were using much larger pieces of timber.

It's safe to say that during the time I was working opposite Hill, I had a fraction of that kind of ability to blast a perfect round. I was keenly aware of my limitations, and it ticked me off because I was good with a jackleg drill and would have preferred to stand with the machine all day as opposed to lifting hundreds of pounds of timber and running

Underground and Radioactive

around the mine begging to have supplies hauled to me or doing so myself.

Had I been doing the majority of the drilling, there is no doubt we would have had some unexploded powder left in rounds, in which case Hill would have been displeased and I'd have been drilling them out. Such a scenario would have made my prospects for long-term survival poor, so I decided to time things so Hill coincidentally did most of the drilling.

I had watched Cal drill and load many rounds but had very seldom ever done so myself. There is a precise pattern that must be drilled, into which powder must be loaded, and timed in such a way that the ore you were after ended up in a big pile of muck. Ideally, what you would be left with was a relatively square, open area in which to set ground support timber. Conversely, setting all the powder off simultaneously didn't do a lot but make a big noise and a small pile of muck.

Individual sets of holes filled with powder had to be blown using timed-delay fuses. When one set of holes blew, it created pockets for each succeeding blast to blow into. Drilling is an art form, and to be good meant mastering the subtleties. I knew how to drill well, and I loved drilling, but I was not good at sequencing patterns.

We used two variations of the Ingersoll-Rand rock drill at Section 35. One we called the 300 and the other less powerful one the 250, but more often we just called them machines. The drill was powered by compressed air. A drilling steel with a hardened bit on the end was inserted into the drill. The combination we simply called a steel.

There was a hole drilled through the center of the steel-and-bit combination through which water was pumped during drilling to eliminate rock dust. An oiler device about the size of a soda can was attached to the compressed air line that kept the drill lubricated. When the drill was running, it produced its own unique smell of compressed air, rock, water, and oil. I found it very pleasant to be around and have never experienced it elsewhere.

The compressed air that powered machines of all types underground was piped down from the surface. The air pressure was very strong, so it was imperative that compressed air hoses have a safety lanyard attached to the hose and whatever machine you were operating

I preferred the larger Ingersoll-Rand machines but this is an example of one of the slightly smaller Gardner Denver drills we used. The fitting in the forefront on the right is for compressed air and oil (with a small rope safety lanyard attached), while the smaller fitting just below and in front of air fitting is a water line. The water ran out the middle of the drill bit and pretty well eliminated rock dust and particulates from the air. These drills were splendidly efficient, rugged and easy to use (photograph by R.D. Saunders from an exhibit courtesy New Mexico Mining Museum).

to prevent the hose from whipping around the area you happened to be working in. A loose compressed air hose could be deadly.

The drill was placed on top of a single pneumatic leg, again operated by compressed air. There was one lever on the machine used to control the speed of the drill and one knob that controlled the height of the leg. Although the combined weight of the drill and leg was quite heavy, they could not be wrestled with and instead required a surprisingly light touch to successfully operate. Helpers new to mining commonly attempted to force the drill this way and that or up and down, and although funny to watch, that method never worked. It was exhausting trying to drill that way. You held them loosely and let the drill do the work.

I became quite good at lifting the drill with the pneumatic leg to the desired height. Slowly starting the machine with one hand while guiding the steel to the desired location with the other was the way to begin a hole. Once it was started, it was "let the fun begin," drilling at full speed with me standing there as relaxed as can be.

The machines were incredibly loud, so another piece of personal safety equipment everyone carried was ear plugs. During orientation we were told to wear them when drilling, but some people never listen. The first time I drilled a round, I was having so much fun I forgot to put in my ear plugs. That rendered me practically deaf afterward, and it took two days for my hearing to return to normal. I'm lucky no serious damage was done, and it was the one and only time I tried drilling without ear plugs.

As was the case with drilling, loading rounds with powder was an art form requiring knowledge of the exact explosive power of the dynamite in relation to the rock at the face and the depth of the drilled holes. Another acquired skill was timing a round with delayed fuses so that different parts of the round went off separately.

The depth of drilled rounds would vary somewhat depending on need, but in a timber stope we typically wanted to cut a near-perfect square. Any more than that and we would end up working beneath unsupported ground. I don't believe such precise drilling was required in a pillar stope although, having only worked with one for two weeks, I would hate to underestimate the skill of a pillar-stope miner.

Several sticks of powder were loaded into the drilled holes with a

Alone

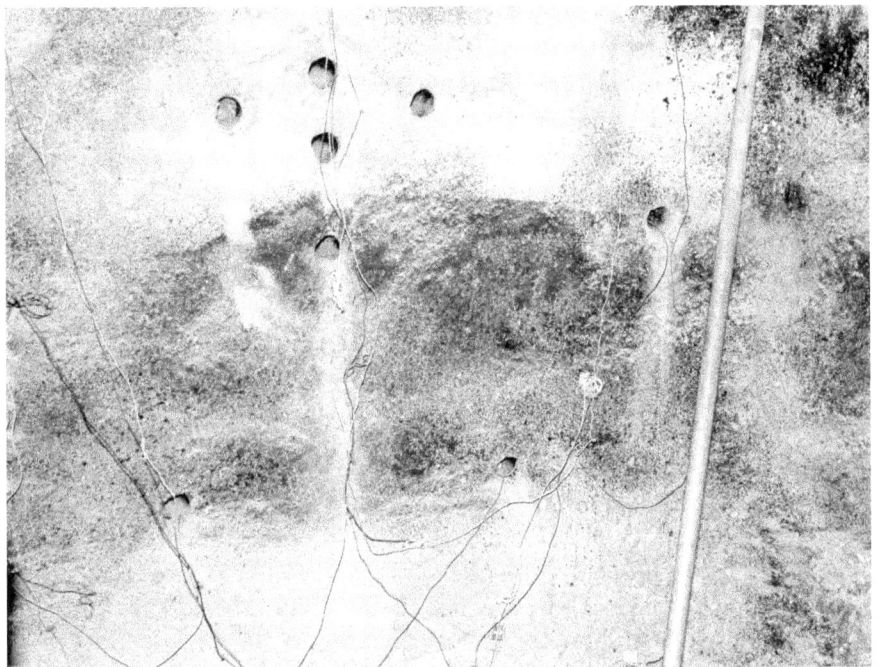

Part of a drilled round with electric blasting caps wired and ready. The pole to the right is a wooden tamping stick used to load powder into the drilled holes. The wooden pole eliminated any chance of sparking that could prematurely and catastrophically set off the dynamite (photograph by R.D. Saunders from an exhibit courtesy New Mexico Mining Museum).

blasting cap inserted into the last stick in each hole with an attached fuse. The powder was then tamped with a wooden loading stick. Using a steel to tamp powder with a blasting cap in place could easily set off an explosion. Unbelievably I had seen this done, thankfully without the explosion, by Riordan while I was working with him on the main track drift.

When you were not illegally hand lighting a round, the blasting caps had electrical connections instead of traditional flame-lit fuses that had to be tied together in such a way so as to time the blast of each hole. Each electrical connector was color coded, indicating what the timed delay was. When a round went off, it sounded like a short burst from a gigantic machine gun.

After a blast what was left was a large pile of muck. The muck had to be taken from the face to the ore chute using a machine called a slusher. The slushers we used were large, motorized drums with heavy cable spooled around them. In fact, they looked like a gigantic spool of thread. There were two cable ends extending from the spool, one for dragging the bucket forward and the other in reverse.

The miner ran the two cable ends out to the muck pile. Standing on the muck pile, he drilled a hole into the face, inserted a rock bolt, and attached a block and tackle to the rock bolt. One end of the cable was run through the block, and then both ends of the cable were attached to a large bucket, one in the front and one in the rear, that was used to drag it through the muck.

The miner got behind the slusher, which was always placed just behind the grizzly, and by manipulating two levers ran the bucket back and forth, pulling it over the ore chute. Slushers were electric and very powerful little machines. There was a wire screen in front of the slusher operator that protected the operator from a whiplashing cable that snapped or separated due to a poor cable splice. I never experienced a cable snap while running a slusher, but I did see a few very poor splice jobs come apart. Having heard stories of missing body parts caused by broken cable, I made sure the protective screen was always in place at least half the time that I ran a slusher.

Cable splicing was another art form, and most miners I knew were proficient at it. Once learned, splicing meant never needing a cable clamp again. As a helper I sometimes looked for broken cable just to practice on. Now on those rare occasions when someone says they need a cable clamp for something, I'm in heaven explaining and demonstrating how to splice cable.

I was fairly proficient at most aspects of timber stope mining with the exception of those pesky timed patterns, so in working opposite Hill, I felt I could keep up with him by erecting square-sets most of the time while he did the drilling.

Hill would drill and blast a round, and I would come in to set the timber. That went on for several weeks. I was happy and Hill was happy. Anyone looking at the contract rates posted in the Dry would think we had a good thing going.

As I arrived in the lunchroom for my swing shift one afternoon,

Slusher used to drag a bucket full of ore from the face to the grizzly. The horizontal bars in front helped to protect the operator, standing to the rear, from the whipping action of a snapped cable. Ideally there would also be sturdy wire mesh installed over the bars to further reduce the possibility of injury. The item hanging on the post to the right just below the KEEP OUT sign is a cable cutter, a handy item often used to cut off the frayed ends of the snapped cables (photograph by R.D. Saunders from an exhibit courtesy New Mexico Mining Museum).

Bill told me that Hill hadn't made it in for day shift. This was bad in that it meant I would have to drill and time a pattern on my own.

I was more than a little concerned the blasting sequence would get screwed up and I would either leave unexploded rounds or have a partial blast that would have to be redrilled, angering Hill to no end. Then the solution came to me, inspired by my experience getting the stuck ore chute open: More powder is better. I would overcome any error in timing the round by adding powder, a lot more powder.

That was a good shift. I had more fun drilling than I had had for

Underground and Radioactive

weeks. Although I hadn't drilled for a while, I had no trouble picking it up again. I drilled the round out in what I thought was the proper pattern and was feeling good. It looked good anyway.

We would normally use four sticks of powder per hole, each stick being about a foot long. I decided to add a little powder to each hole in the hope of forcing the ore to do what I wanted it to do should I somehow screw up the timing sequence of the blast. A likely scenario.

For my eight-foot round, I used six sticks of powder per hole times sixteen holes in the pattern for ninety-six sticks of powder, give or take a stick. That's a lot of powder and far more than should have been used in a normal round. Then I added the timed fuses, not really certain what order to use.

I timed my round by hooking up the blasting cap wires in the sequence that looked right and tied the whole thing into the central blasting network. No hand-lit round here, because when that baby went off, I didn't want to be anywhere near it.

By the time the sequence was all tied in, the shift was over, so Art and I headed back to the lunchroom, where, after everyone had been accounted for, Bill, using the electric blasting board, would blast all the rounds for the day. With the shift now over, we returned to the surface.

As I was changing in the Dry, I kept an eye out for Bill, half expecting him to come looking for me. Only the bosses would have been around when the rounds were set off, and there was the possibility that my extraordinary round had been, if not discovered then, heard even in the lunchroom. If so it would have left Bill and all the bosses wondering what the hell had just happened.

Bill never came looking for me, though, and, hoping against hope that my extra powder round had worked, I left for home.

The following day I came into the Dry and saw Hill already dressing, having just come off his shift. I didn't have to go over to him because he came over to me. Oh shit, I thought. But suddenly a big smile broke out on his face. He said he couldn't believe how much we had done the night before. OK, evidently my mega-round had worked well. Using my best "Well, Hill, we do what we can do" expression, I explained that yes, we had put in a good shift, as if to say, "What did you expect?"

Alone

At any rate Hill was very impressed that day. He told me they had spent the whole shift slushing and getting the chute pulled and still hadn't gotten all the ore out.

While I could never get a motorman when I needed one, Hill had no problem doing so for reasons that I never fully understood. Maybe he was bribing motormen, but the chute kept getting pulled all day until they had sixty or more carloads with plenty more to pull.

When I saw what had happened in the stope, I could hardly believe it myself. The round had blasted through fourteen feet lengthwise and eight feet up. Not only that, but we had broken through into another timber stope not being mined at the time. Our huge blast had created one monumental muck pile and one monstrous, wide-open ballroom.

Hill said I would have to keep on filling the chute to make any space to set timber so we did. All night we filled that chute, getting another forty cars of ore out of there. That week we moved way up on the contract rate board in the Dry.

We had also created that ballroom, and it was a big, big hole. Breaking into the other stope had created some problems with ground support as well. We had not only had a giant hole in our own place but had caused some massive cave-ins in the other empty stope. All of that open space had to be supported somehow, so we needed timber and a lot of it.

Ballrooms were no fun to work in but sometimes a spectacular sight to see. They were unsupported ground above our heads from twenty to as high as a hundred feet. When I ran into a few of the larger ones during my wandering laborer days, I would be awestruck by the view.

Ballrooms were unpredictable. Occasionally they would sit empty for long periods of time and nothing happened, while in others slabs large and small regularly dropped from the back. When a monster slab fell, it would do so with a shuddering thump. The one thing every ballroom had in common was if it was not supported in some way, it would continue to cave in.

In our case we had several huge boulders of ore sitting out on the ballroom floor that we had to break up and send down the chute two hundred feet away. The only way to do that was by using powder.

If a slab was small enough, a stick or two of powder jammed

Underground and Radioactive

beneath it or in a crevice would do the trick but for larger, slabs ten feet high or higher, that method had little effect. The only way to break them up was to drill a hole or two, pack them with powder, and blast—not a safe or easy task.

What I did next was probably the most dangerous work I ever intentionally undertook at Section 35.

I devised a plan whereby I would set up a drill in a relatively safe area with enough slack in the water and air hoses to reach the target boulder field on the ballroom floor. I estimated the distance to be around fifty to sixty feet. When everything was set, I would run out into the ballroom with the machine in hand, hoses dragging behind, and drill a six-foot hole in the boulder as quickly as possible, then run back for the safety of supported ground.

Those machines made an incredible racket, and my biggest fear was that the reverberations would kick something loose from above that could very well land on me. It was not an unreasonable fear, seeing as small slabs were falling regularly.

Art and I went ahead and readied the machine for drilling, making certain there was plenty of slack in the hoses to reach the designated boulder. I stood at the edge of the ballroom for quite some time examining the situation, eyeballing the exact location on the slab where I would drill. Very, very seldom during my time underground was I ever nervous, but standing there I was.

Knowing that thinking underground never came to much good, I double-checked that the air and water that the machine needed to operate were both on, then took off running across the ballroom for the largest boulder. When I came to it, I slapped that machine on full, guiding the steel by hand and steadying it with as much force as I could muster, and drilled a nice six-foot hole.

It was very fortunate that the slab was rather high-grade uranium ore, which tended to be fairly soft rock and easy to drill into. It took less than five minutes to complete, but it seemed a lot longer than that to me, being as exposed as I was.

Thankfully nothing came down from the back while I was drilling. Had anything been ready to fall, the blast I was about to set off would shake it loose. I readied enough powder to fill that six-foot hole and shoved a blasting cap and a fuse to hand light in it. Again sprinting out

Alone

to the boulder, I rammed that powder into the hole using a tamping stick, lit the fuse, and got back out as fast as I could. With a lead time of three minutes or so, Art and I had plenty of time to get down the manway and wait for the explosive report at track level.

The resulting blast obliterated the target boulder with nothing further falling from the back. One down and several to go.

Over the course of the next few hours, I repeated the same process for the remaining boulders. Almost miraculously, not only did nothing fall on or near me while I was drilling, but nothing significant at all ever fell from the back as a result of the blasts. Sometimes you just get lucky.

The final piece to the puzzle was setting up a very long slusher cable, which required drilling another hole out in the ballroom against the far wall, into which a rock bolt would be inserted with a block attached. I did that too, and it was the final time I ever pressed my luck working in a ballroom.

Blasting all those boulders got us many more cars of good ore that took well over two shifts to slush out. When we were done, what was left was a huge open area with a relatively flat floor ready to timber.

For the better part of two weeks, Hill and his helper and Art and I worked building square-sets. Not only was I good at erecting sets, but it got to the point where I was able to pull ahead of Hill. If he put up four sets a night, I would put up six. I felt pretty good for a while there.

My experience with Cal had taught me how to build good, level ground support, but Hill, although good, was not the perfectionist that I was. I'd notice some uneven work of his when I came on shift that needed fixing on occasion—no way anything other than plumb and level was going up in our stope.

Square-sets paid well, and we continued to make good money. I was up around the fifty-dollars-an-hour mark during that period and just entering the fringe of good miner territory on the contract board—a very satisfying feeling. It was the only period I managed to do it with any consistency.

After my fortuitous big blast and about four or five weeks of big money, things settled back into a routine. Hill again did all the drilling and blasting, and Art and I would muck it up, with me sometimes running the motor and pulling the chute.

As good as Hill was, I discovered that he had worked several mines in the Ambrosia Lake area, becoming somewhat of what we called a tramp, but not in the vagrant sense of the word. For whatever reason they tramped from one mining company to another, I suppose looking for greener pastures. Tramp miners, or helpers and laborers, for that matter, generally knew a lot more than someone like me, as they had seen a variety of ways other mining companies operated.

Section 35 seemed to attract a lot of tramp miners because of the exceptionally fair contract system that was used. Again, not all mining companies paid contract rates in place of a flat hourly rate. Having seen how much of an incentive contract mining could be, I always wondered how other hourly rate companies kept their production up, or their best miners, for that matter.

Hill didn't hang around any one place for long, apparently, and one day he vanished without a word to me. There was a rumor he had moved on to uranium mining in Green River, Wyoming, but nobody knew. Shotgun immediately assigned Bill Davis to work opposite Art and me.

Davis was another of Cal's former helpers and another very good miner. Apparently he had learned a lot more about mining from Cal than I had and had become a top earner. Davis knew Cal had liked me and knew I was good at a few things, so we got along just fine. Soon we got into a routine similar to what Hill and I had. Bill would drill and blast, and I would muck and build square-sets. I'm not sure that either Hill or Davis knew what was happening by my always ensuring they would drill and I would timber, but it kept on working, and everyone was happy.

Davis and I had been working opposite shifts for a few weeks when Mel called me into his office and told me from now on I would be mining the stope alone and working twelve-hour shifts.

I had heard of a few miners working twelve-hour shifts in stopes that were high-priority, high-grade ore areas. In those stopes each shift of a miner and helper would work twelve hours on and twelve hours off, keeping the stope producing twenty-four hours a day. I had never seen or heard of an average stope having a single crew working twelve-

Alone

hour shifts with no opposite crew. From now on I would be working from 4:00 a.m. to 4:00 p.m. each day.

Not being too thrilled about the twelve-hour shifts was one thing I was able to overlook fairly easily. I was really pleased at being able to work alone with no opposite crew to answer to and even more excited at the prospect of a bigger paycheck by working those sixty-hour weeks.

True, it was very unusual but not unheard of for a miner and a helper to work an average stope without a crew working opposite them, but Cal had always done it. If Cal could it then I could do it, I thought.

Bill was also assigning me a new helper, Jack Stutts, who I knew peripherally from having worked with him occasionally early on when I had been assigned to the station loading and unloading supplies. Not a bad worker but not a go-getter, and he complained.

Some of the work we did was painful, and one surefire way to make it worse was listening to complaints from Stutts. The entire day was about how much he was hurting and how little we were being paid. I had no such complaints considering what I had been getting paid prior to hiring on at Kermac, but thanks for reminding me about the pain, Jack.

Evidently Stutts had finally moved up and was now a helper. He was no kid, having been underground for quite some time, though without making much progress. Additionally, I came to find out that he had recently been assigned to several miners, each of whom had unloaded his complaining butt.

Stutts was otherwise a likeable enough guy in his late twenties. He was very thin, was about five foot eight, and didn't appear to be able to do the kind of heavy lifting required in a timber stope. I was worried he wouldn't be able to either lift or reach up high enough to set eight, ten, and twelve-foot posts for our square-sets. I need not have been concerned; he was one strong little guy who could do the work.

Being somewhat of a tramp himself, Stutts had worked in several mines as a helper over the years and for many miners in a variety of stopes. Frankly, he seemed to know more about mining than I did. He talked a good game, and that was the problem: talk.

I found that in addition to complaining, Stutts talked a little too much, and he asked too many questions. I got especially annoyed when he asked what we were doing, because I didn't know a lot myself at the time. I ignored most his inquiries. I felt right away that he wouldn't last long with me.

A lot of people, me included, were called pard rather than by first or last name, but some people, including my new helper, were known by their last names only. I would have preferred pard, but I sensed he seemed to prefer being called Stutts, so I conformed.

Twelve-hour shifts proved to be difficult. During an eight-hour shift, we would work about six hours maximum. The other two hours were spent getting dressed, going down and up in the cage, and walking back to the stope. During twelve-hour shifts, we were working ten hours.

It was life in the dark during those shifts. We would report for work at 4:00 a.m. and get off at 4:00 p.m. During the winter months, it was already dark by the time I exited the mine after a shift, and there were times I would barely see the sun at all.

No extra laborers were showing up at 4:00 a.m., so if Stutts didn't report, I couldn't work for at least four hours or until the day shift arrived and I could be assigned a temporary helper. Thankfully Stutts was good about coming to work, but because we could never work alone, there were some days I sat in the lunchroom fuming for four hours waiting for a helper.

Our production was surprisingly good. We averaged around twenty-five dollars an hour. Staying true to what I had learned from Cal, I gave my helpers four hours of contract time on eight-hour shifts and six on twelve-hour shifts.

I learned more about drilling, blasting and timing the patterns working alone. When I made a mistake, I was accountable only to myself to fix it, and only I paid for it in the way of reduced contract rates. I had settled in to being a lead miner.

Darkness

Being underground without light meant perfect darkness. There were no reflections, no glints, no shadows. There was only blackness. I could hold my hand a fraction of an inch in front of my face and see nothing. It was the essence of black.

Sometimes I would turn my light off to stand in total darkness. I found that disorienting to the point where I would lose my equilibrium, becoming unsure of whether I was standing straight up or leaning over. It was a very odd feeling.

I spent a good deal of whatever downtime I had sitting in the dark, listening. I noticed that sounds seemed to be amplified. I could hear the motors rumbling on the track below, ventilation fans humming, and, faintly, the far-off sound of a machine drilling in another stope. Occasionally there would be creaking from stulls or square-sets as the pressure of the earth slowly compressed them and the sound of a small slab falling somewhere nearby. Fascinating stuff, but only if being in the darkness was voluntary.

As happened to almost everyone at one time or another, I inadvertently picked up a partially charged battery from time to time, and my lamp would go out during a shift. Most often there was a partner around to go fetch another battery, but not always. In that case I found myself sitting in the dark, waiting for someone to wander through.

Without an alternative light source and alone, it was best to sit and wait to be found should your lamp go out. Usually a partner would show up or a shift boss would amble by, but no experienced miner, partner, or laborer was ever without light unless it was planned. And no miner, helper, or laborer I knew ever sat and waited for anything. That was why everyone carried a disposable butane lighter.

On the surface a lighter is a tiny, insignificant source of light, but

Underground and Radioactive

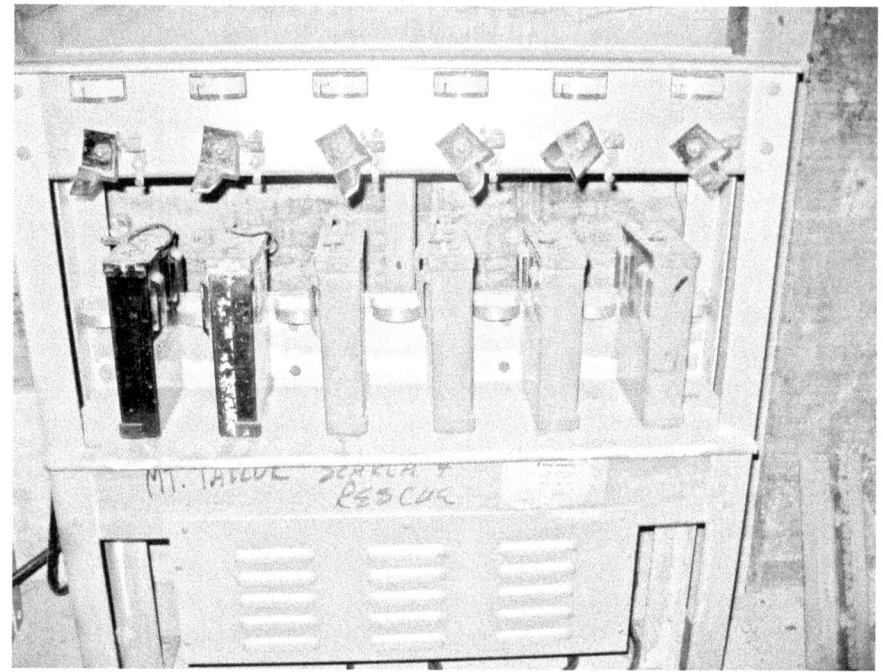

Rack for charging batteries. The gauges visible at the top of the rack above each battery were supposed to indicate the level of charge. These gauges were all too frequently wrong, resulting in a miner with a dead battery being left in the dark, as happened to me more than once (photograph by R.D. Saunders from an exhibit courtesy New Mexico Mining Museum).

in total darkness it seemed like a blowtorch illuminating a hundred feet of space. Most miners I knew used one to find their way out of a mine when their main lamp went out. A small butane lighter was very light, durable, and reliable, and, unlike a small flashlight, always worked.

Before each shift I would go to the battery rack, where several dozen battery packs were charging. While there were gauges showing the strength of the charge, it was not that unusual to pick up an undercharged unit every now and then despite the gauge showing a full charge.

The battery packs hooked to a miner's belt, and the headlamp attached to the miner's helmet. After hooking it all up, I would usually test the lamp before heading down for my shift, but every now and

Darkness

then I would overlook it and take an undercharged battery with me. Even then if the light started getting dim, there was usually time to go back up for another, but not every time.

I had been working several months of twelve-hour shifts and was getting run down. I had been through a bad cold and a bout with strep throat and was overall fatigued, so I wasn't surprised when I picked up an undercharged battery and found myself with a very weak light one morning.

Occasionally, if Stutts hadn't shown up and was late, I would head back to the stope. I would slip out, so Bill usually didn't notice, and Stutts ordinarily showed up anyway. If he didn't appear in a reasonable amount of time, I would go back to the lunchroom to wait it out.

Of course on the day I picked up the undercharged battery, Stutts failed to show up, and just as I was stewing over what had happened to Stutts and contemplating returning to the lunchroom, my battery went dead, leaving me alone in total darkness.

I sat down and reached for my butane lighter that I usually carried in one of the upper pockets of my overalls and found it missing, so there I was, unprepared, in total blackout, with no way to move safely in any direction. This left me with one option: waiting for the shift boss to make it around sometime that morning. So I sat down and waited.

Hours went by, and while I hadn't panicked, I was certainly ill at ease. I could well be waiting there until the shift change when the boss would check the in-and-out board and realize I had not returned. He would then either look for me himself or a search party would be sent out and I would be found. But if the boss didn't notice I was a no-show and had not checked the in-and-out board, I might be sitting there until 4:00 p.m. Kermac bosses were fairly observant of the most important safety measures like the in-and-out board, so I was confident in being found eventually, but it was frustrating sitting there in the dark. It was no longer a game.

Finally, after what I estimated to have been about four hours, I started to see some light coming from the manway and heard steps coming up the ladder. Must be the shift boss, I thought. Sure enough, a boss from day shift, Bill Purcell, had noticed me missing and had started searching on his own. Thank you, Bill. It was an interesting few hours and taught me a little bit about what total darkness really is.

Ghost in the Stope

In mining, there are all sorts of mechanical sounds, from those muffled, rumbling explosions that I loved to hear to the sounds of the shifting earth and creaking timber that I didn't like so much.

Smells were also unique to underground mining. The relatively confined spaces could have contributed to that, but I never again experienced those smells after I left underground mining.

My favorite smell was the one associated with drilling. The combination of compressed air, water, heat, and oil was special. I never heard anyone else mention it or thought to bring it up myself, so maybe it was just me, but I loved it.

Most of the machines at Section 35—muckers, drills of all kinds, and chippers—were run by compressed air, and they all produced different smells. Different kinds of rock had specific smells too. Uranium itself I could smell and differentiate depending on the grade. Mostly we mined uranium that was close to black. That was the good stuff, but the very high-grade ore was sometimes close to yellow, and the two had a different smell. (As an aside, only the very best miners mined high-grade uranium at Section 35, but every now and then a miner like myself would run into some on a small scale.)

I always especially enjoyed walking down the main track drift as the place where so many odors were combined with the fresh air rushing out of the main shaft.

As I mentioned, total darkness was an experience. So was light. In total darkness any light source brightens a very wide area. I was always amazed by what those portable butane lighters could do in that respect. A match could do the same thing.

It became much easier to imagine how civilizations managed to function at night before the invention of the electric light. Gas lamps

I'm sure worked very well at night, as would oil lamps and candles—or, for that matter, how miners worked for many years using carbide headlamps and before that candles.

It doesn't sound like much, but the miner's lamp that we had attached to our hardhats was a major technical advance in mining. In total darkness, it easily lit up a very large area. Combined, a miner's and a helper's lamps easily lit up the immediate area they were working in.

Any light source could be a seen a long way off. That's why the relatively low-powered trip light on the back of an ore car was so effective. In otherwise total blackness, you could see that thing coming toward you with no problem.

Light was also used as a communications device underground. If you were too far away from someone or if it was too noisy to hear them, the headlamp was used to send signals. For example, moving your headlamp up and down meant to back away or go away, and a circular motion meant come forward.

An individual walking down the main track drift two hundred or more yards away was very easily seen, so we always knew when someone was around or approaching. There was no chance for a shift boss to sneak up on us as we worked in a stope because we could see his head lamp and even the glow of it as he climbed the manway. (As a rule, incidentally, I never liked the shift boss to come around. Nothing personal, but a shift boss visit cost money because it meant stopping whatever I was doing.)

The 812 was developing into a stope with depth but not much width. It couldn't have been more than twenty yards across, and there were now three stories of eight-foot square-sets. We were working well over five hundred feet from the manway entrance.

One day I was drilling a round when Stutts motioned to me that there was light down by the manway. It wasn't the usual time for Bill to come around for a visit, but obviously someone was on the way. I stopped drilling and with my headlamp motioned the visitor, who had stopped, to come forward.

Whoever it was stayed put. I thought that was unusual, so I told Stutts, "I'll go check it out."

As I started back toward the manway, weaving between the posts of the square-sets, the light vanished. That was odd.

Underground and Radioactive

Continuing on to the manway, I saw no light down the raise and no light anywhere else, for that matter. I should have at least seen a glow. Again, that was odd, but stranger things had happened, so I turned around and walked back up to the face to finish drilling the round.

When I got back to the face, Stutts asked me, "Who was that?"

"I don't know. I didn't see anyone, and nobody was in the raise."

"It had to have been somebody," said Stutts.

"I know, but there wasn't anyone around, so whoever it was must have been in a big hurry to get out of here."

We went back to work drilling, with Stutts looking puzzled and me happy not to have been bothered by whoever it had been.

About an hour later, I was just about finished drilling when Stutts again motioned to me that there was a light down by the manway. Getting annoyed, I again stopped drilling and headed back toward the manway. As before, the light vanished.

This time, determined to see who was visiting us, I used my previous expert experience of getting down manways quickly and descended to track level. When I got to the track, there was no visible light anywhere to be seen.

I even turned my light off, knowing that the pitch black would show the light of any other person no matter how far down the drift they were. Nothing.

Well, this is a little odd, I thought.

Just then the warning lights on the back started flashing to indicate that a motor was coming my way.

I ducked back into the manway as the ore cars started passing by. When I saw the motorman, I motioned with my head lamp to stop, and he did.

"Did you see anyone walking along the drift?" I asked.

"Nope, didn't see anyone" was the motorman's response.

"Somebody was up in the stope just now. You haven't seen a boss wandering around?"

"Nope, but he might have been in a cutout and I missed him, or whoever it was."

"Yeah, OK. Probably. Thanks."

The motor continued on, and I climbed up the manway and back to the face, where Stutts, as before, wanted to know who it was.

Ghost in the Stope

"Who was it?" he asked.

"I didn't find anyone, and I went down to the track."

"That's impossible," he said.

"You would think so, but I'm telling you nobody was there."

Stutts seemed nervous at first and then acted as if he thought I was messing with him. Had I liked Stutts more, messing with him would have been a possibility, but in this case I didn't and it wasn't. He was a fairly good helper but didn't have a great sense of humor.

I figured it had to be a shift boss, a lost laborer, or a geologist. True, I hadn't found whoever it was, but it was somebody. No doubt about that. That was the end of the excitement for the day, and soon the shift ended.

The following morning at 4:00, we were back at it. This time we were drilling a hole for the rock bolt to install the block for slushing out our muck pile. In the middle of drilling the hole, Stutts motioned that we again had another visitor down by the manway. Now this was very unusual, because it was an odd hour and unlikely that a shift boss or geologist would be roaming around.

Very annoyed this time, I told Stutts I wasn't going to check it out again and that whoever it was could come on down to see us if he wanted to, but I wasn't going back there again.

Stutts said he would go look this time, and that was fine with me. I watched him head back and saw his lamp, but the other light again vanished. Then Stutts vanished as he bolted down the manway to catch up with whoever it was.

A short while later I saw the glow of a light come up the manway and then a brighter light head toward me. It was Stutts, who reported that he couldn't find anyone in the stope or on the track drift.

I could see Stutts was really doing some thinking. So was I, and my thoughts were of what I'd been told many times by Cal: it just doesn't pay to be doing too much thinking. It was then that it dawned on me that maybe the reason Stutts was still a helper after so many years was that he had been thinking.

We continued to work that morning until around eight. That was when the light again appeared. I had had it with the light and ignored it this time. Whoever it was could come on down and see us or not. If they, or it, wanted to play games, I didn't want any part of it, and they could just keep going away or disappearing in whatever way they had managed to avoid us. But poor Stutts was really getting excited.

Running back down the stope at full speed, Stutts was determined to get to the bottom of it. Again, I saw the light vanish, and again Stutts went down the manway looking. When he came back he appeared to be shaking and talking about how there had to have been somebody there, and where did they go? I didn't know and didn't care, but evidently Stutts did.

We didn't see the light again that shift, but poor Stutts was not much use to me the rest of the day, nervously looking around and not seeming to be interested in much else other than the mysterious light. That was the last day Stutts worked as my helper.

After the shift Stutts told me he was not going to work with me anymore. "That fuckin' place is haunted," he told me.

"Well, if it is, whoever it is doesn't mess with us, so who cares?" I replied.

"I care, and I ain't workin' no fuckin' haunted stope."

In the cage on the way to the surface, Stutts told me he was going to see Shotgun and ask to be taken out of the 812. Thinking that might be a good one to see, I decided to tag along.

Explaining the light we had seen to Shotgun, Stutts began talking about how the stope was haunted. I don't know if Shotgun thought I had done something to convince Stutts there were some strange things happening in the stope, but with all his experience underground, I doubted it. I thought maybe Shotgun had even seen something like what was happening in the 812 before.

When Stutts was done jabbering, Shotgun had me explain the light. I told him that, yes, it had been there but kept vanishing, just as Stutts said. I went on to say that whatever it was did not bother me and was not hampering my work, adding that I just didn't care about whatever it was causing the light.

Frustrated, I didn't care that Stutts was standing there, so I told Shotgun, "Stutts is fucked up, and he's not helping get the work done, so if he wants out, it's OK with me." That was enough. Stutts was taken out of the 812.

The following morning, I was introduced to my new helper, Daniel Ortiz. I had worked with Daniel before on some small jobs around the station the year before. He was a nice guy, diminutive appearing but deceptively strong with a good attitude. What I appreciated most was that he just did the work and didn't talk very much.

Daniel had moved up to miner's helper in good time but, like Stutts, had already worked with two other miners in the past few months. A pattern was developing whereby I sure did seem to be getting the helpers who maybe nobody else wanted.

Why whoever had made the decision thought Daniel would be able to handle the work in a timber stope where we were building eight- to twelve-foot square-sets is another mystery I'll never figure out. Normally the shorter guys were working pillar stopes, where finesse and knowledge matter a lot more than physical strength.

I knew immediately that whatever future Daniel had working underground, it was not going to be in a timber stope. He was willing to try and a hard worker but just not able to consistently lift heavy timber, as I soon discovered. It also seemed he had a little bit of a spiritual streak in him.

As Daniel and I walked back to the stope, I wasn't thinking about the light and how Daniel might react to it. We had a few sets to build that day, and I was wondering how, Daniel being such a short guy, would we be able to set the eight-foot-long cap timber on eight-foot-high posts?

Soon we were hard at it, measuring and cutting posts and caps. Sure enough, when it came time to set the caps, Daniel couldn't lift them high enough. I ended up balancing my end on top of a post while Daniel held his end about chest high. After my end was set, I would take Daniel's end and lift it up into place. That system worked, but it was going to be very difficult to work that way unless someone figured something out. To his credit Daniel did just that.

Eventually Daniel took to setting a twelve-by-twelve scrap of lumber on his end to stand on so he could set the cap himself. That system worked well. Then the light returned.

Sometimes, and it usually happened early in the shift, I would glance down through the sets toward the manway, and there again was the mysterious light. I never brought it up to Daniel, but eventually he did see it. Here we go again, I thought.

The first time Daniel saw it, he wondered aloud who it might be. "Who the fuck is that?"

Underground and Radioactive

I told him I had seen the light many times and had checked it out many times, but that there was nobody there. That might not have been the best approach.

Confused, Daniel wanted to know what that meant. "There has to be somebody there," he said.

I explained what had been happening and that each time we went down to check we found there was nobody. Of course Daniel didn't think that was possible, so I told him to go take a look if he wanted to. He did and proceeded down the long row of sets. The light then, as always, vanished, and, as had Stutts, so did Daniel, right down the manway.

Soon he was back, looking confused and plainly worried. Seeing his expression, I explained how the light kept showing up and how we had never found anyone in the stope. For my part I wanted to get on with the work and ignore the light. Whatever it was wasn't bothering us and left when we went to investigate. So why bother with it?

I did my best to explain to Daniel how long the light had been showing up, that there really was nobody there, and not to worry about it. That didn't work very well either.

From then on Daniel was uneasy and, just as Stutts had done, continued to look for the light to show up instead of concentrating on what we were supposed to be doing. The light appeared at least twice more that day, and each time Daniel went to see who it was, and each time he found nothing.

At the end of the shift, he was even more shaken than Stutts and said the stope was haunted and he wouldn't work in it again.

"What makes you think the stope is haunted?" I said.

"You ever see a light like that before down here?" he replied.

"No, but why does it matter? It never comes close to us and never bothers us."

"It bothers me" was the last thing he said.

When he hit the surface, he went straight to Shotgun, told him what had happened, and asked to be removed from the stope.

That was good news for me. First, it confirmed for Shotgun what Stutts had said previously. I couldn't be blamed for doing something to either one of them. Second, I would get a new helper who might be able to lift timber more easily.

Ghost in the Stope

As he had with Stutts, Shotgun, always a very fair guy, removed Daniel from the stope. He had lasted exactly one day.

After that shift I finally gave some thought to the light and figured whatever it was or whoever it was wasn't interested in talking or interrupting our work. If it was a ghost, as Daniel and Stutts thought, then it wasn't doing any harm and was only watching. That was fine with me; however, someone wanted some answers.

When I reported for my next shift, I had been assigned yet another helper, Tom James. Also appointed to us was our own geologist, who was to spend the day in the stope. His assignment was to stay with us and if the light appeared to get to the bottom of it.

"Who told you to hang out in the stope?" I asked.

"Listen, I'm just doing what I was told to do," he said.

"Yeah, OK," I replied, but I was sure suspicious of the idea of having someone hanging around the stope while we were trying to get some work done.

Why a geologist and why whoever had assigned him to that useless task cared at all about a mysterious light always remained unknown to me. I certainly was never on solid enough ground with either Shotgun or Mel to be questioning their decisions. The directive might not have come from either of them anyway, but I guess I will never know. Carry on with the work was my attitude as the three of us headed back to the stope, where Tom and I had more timber to set and the geologist had some sitting around to do.

I don't recall who the geologist was, but I knew and liked most of them. They were usually recent college graduates, and working for Kermac was their first job. Their most distinctive characteristic was they were always clean: clean hands and face and clean clothes with tucked-in shirts. That look really made them stand out underground, where everyone else was filthy.

The geologists carried around long or short probes that they would insert into exploratory holes drilled by the long-hole miners or muck piles of recently blasted ore. Long-hole drills were much larger than those we used in production and looked to me like a lot of fun to operate.

Unlike jackleg machines, a long-hole drill was stationary and well braced against the back and floor. Other than having to carry them

Geologist's probe used to test ore quality. This particular probe type was plunged into a recently blasted pile of muck to test the grade of the ore (photograph by R.D. Saunders from an exhibit courtesy New Mexico Mining Museum).

around, set them up, and add lengths of steel, running a long hole wasn't a bad job. It did take a skilled miner to do it, and it paid well. Unfortunately, I never had the opportunity to run one.

The long-hole miner on the 1–5 was a guy by the name of Botts. He was recognizable by the yellow rain suit he wore all the time. That was unusual because the 1–5 was above the water table and was very dry. Had he been working the wet lower level, I could have understood his wearing the suit. Below the water table, water came off the back so much it was like rain, making a rain suit mandatory.

We had no long holes drilled in our stope, but the geologist had brought his short-hole probe and did use his time to look around the place and probe a few small muck piles checking for ore quality.

While the geologist was probing close by, James and I resumed square-set construction. James was a good-sized guy, and he had no problem setting caps and getting our sets caught up to where we could start drilling again. Another guy who had some experience in several mines, James knew how to set up the drill. He was busy doing that while I was laboring at the face, trying to figure out what kind of a pattern I needed to drill.

All three of us saw the light when it appeared as usual down by the manway. James of course wanted to check it out, and I once again said it was nobody. "There's nobody there, and there never has been with that damn light, so let's just get on with what we have to do," I said.

That kept James with me and working, but the geologist was convinced it had to be somebody, so he was the one who took off back toward the manway. "Nope, not this time. That's somebody for sure," he said, scurrying toward the light source.

Of course no sooner had he left to investigate than the light vanished. Not satisfied, he climbed down the manway to the track drift and, as with all of us before him, found nothing.

The light appeared at least two more times during the shift, both of which the geologist observed. He went to investigate each time and as always found nothing.

James, as with the other helpers, swore he wouldn't work in the stope again. The geologist took a more pragmatic view and attempted to explain the light to us as possibly being static electricity generated by the shifting ground.

"You see, uranium is a metal, and if large slabs of it grate against each other, it could cause static electricity," he explained.

Now he had me laughing. "Oh, bullshit," I said.

"No, it's true; it could happen, and maybe that's what's going on."

I didn't know much about geology, but slabs of shifting uranium ore causing static electricity? I wasn't buying that, and even if it were true, I had never heard of static electricity remaining stationary for extended periods.

After the shift, the geologist, James, and I went to see Shotgun. James explained how he wasn't working in a haunted stope, and the geologist explained his theory of the earth moving, possibly causing static electricity.

I think in the end Shotgun cared as much about whatever the

mysterious light was as I did, which is to say, not at all. It wasn't bothering anyone and, other than causing one helper after another to flee the stope, wasn't hampering production.

If I had been on the hook with Shotgun, I was off it now. I could get back to work, albeit with yet another helper.

So it was that the era of the mysterious light in stope 812 continued on with a new helper being assigned to me, this time one of the best I ever knew.

Ana Maria

It was April, and the weather was changing on the high plains of western New Mexico. The wind begins to blow nonstop around then, and for the next two months, it howled relentlessly twenty-four hours a day.

I did all I could to keep the blowing dust out of my home and car, but nothing seemed to work, as the ultrafine particles blowing off the desert collected in every nook and cranny.

The cacophonous roar of the wind wore on my psyche, slowly driving me to distraction. The area surrounding Grants and Milan is a wonderfully scenic area of New Mexico, but that wind I could have done without.

Going underground was my only respite where, on the 1–5 level anyway, there were only pleasant breezes and uniform temperatures. Spring, summer, fall, or winter, it was always between sixty-five and seventy degrees.

During the windy season, I returned home from work one afternoon, and there waiting in front of my house was Greg Hornaday. He'd been the one who brought me out to New Mexico and to the fine job at Kerr-McGee Section 35.

Some months earlier, Hornaday had left Grants, possibly to return to Illinois for a break or maybe Wyoming to mine. I don't recall which. When he left he hadn't been very specific about his plans, so I might never have really known.

I was well entrenched when he decided to leave, so it wasn't as disappointing as it might have been a few months earlier. He had always been a nice guy with whom I got along well. But being fairly satisfied with my progress at work and having access to many stories, fascinating characters, and more money than I thought I'd ever need, I'd contentedly gone on alone.

Underground and Radioactive

Now Hornaday was back, and it was good to see him, but I was especially happy to have someone to have dinner with on that particularly day and to escape the maddening wind. One way to do that was the Holiday Inn restaurant and bar.

Over the years I've described Grants in any number of ways. Depending in what context the question is asked, I'll talk about how Grants looked like a boomtown in old western movies, with cowboys or miners running wild in the streets, and suggesting automobiles be substituted for horses in the conjured image. If it never got that wild, it still makes for a decent story and a good picture. It also more accurately conveys my and many of my fellow miners' state of mind.

I might also answer that Grants represented the excitement and romance of the old West. There was plenty of excitement, but with a man/woman ratio of about ten to one, there wasn't much romance for young, single guys. That's what made meeting Ana Maria all the more implausible.

The Grants Holiday Inn restaurant and lounge was at the time one of the more popular high-end spots in town, geared more toward visiting mining company officials than miners. Executives from Kerr-McGee, Homestake, Ranchers, United, Philips, and many other companies stayed in the hotel and relaxed in the restaurant and lounge, making it a very busy place and, far different than the saloons I was used to. There was always a conversation going on that more often than not had something to do with mining somewhere in the world.

Greg and I drove over to the Holiday Inn and before dinner took seats at the bar, where I'm sure we discussed something about what he'd been doing and where he'd been, but I don't remember any of that conversation. I probably didn't hear much of what he said anyway because behind the bar was a beautiful bartender, Ana Maria.

It was early, there weren't many people at the bar, and Ana Maria was quite friendly and talkative—a real treat since there hadn't been many women around to talk to for quite some time, it seemed.

A real beauty, Ana Maria had moved with her family from Guadalajara to Grants. Her father, a true patriarch in every sense, had since passed on, leaving his eldest son to run the family business. He and his brothers owned a successful exploration drilling company in Grants and were doing well. Ana Maria, still living at home, was making her own way working in a variety of jobs including the one at the Holiday Inn tending bar.

I was pleasantly surprised that she spoke more than a few words to me that night. I picked up the accent right away, and when she asked who I was, I changed my name to Rogelio on the spot. Having taken some Spanish in high school and college, I felt confident that I could pass myself off as Mexican.

It was ridiculous, I know, but Ana Maria, good natured as she was, went along with the joke, and we had a nice enough conversation so that by the time Hornaday and I had to leave, Ana Maria had agreed to maybe meet me at the stock-car dirt track the following Sunday afternoon. Maybe was good enough for me.

There weren't a lot of things to do for recreation around Grants at the time other than rodeos, enjoying hikes in the nearby mountains and stock-car racing. There was a dirt track just northwest of town where races were run almost about every Sunday during the summer months. The track had little if anything, in the way of spectator seating, so many of us would drive our vehicles up to the fence that ringed the track for a nice up-close view of the racing action.

There were several miners in the area who liked to build and race modified stock-cars, most of which looked more like former demolition derby entrants. Nonetheless, the races at the small, quarter-mile track attracted decent crowds on Sundays.

New Mexico allowed no alcohol sales on Sunday at the time, and a few enterprising souls showed up with cases of beer that they would offer for sale to racing fans. The price for a beer was ridiculously inflated, but people paid it.

I had noticed that many beer-drinking mine workers were very poor planners. Rather than double their beer purchase on Saturday and save half for Sunday, they either drank it all on Saturday or forgot about Sunday altogether. Fortunately for them beer was available on Sunday in spots. For example, there was a business in a small village aptly named Budville about twenty-five miles east of Grants off I-40 that illegally sold beer on Sunday for double the regular price. First-time customers had to be shown the way to a side window on a crumbling adobe structure where the transactions were made.

Not being a beer lover, I only showed up at the track from time to time for the exciting racing and now for the possibility that Ana Maria might appear.

As I sat in my car watching the races and listening to music, my hopes weren't up that she would show. There were plenty of exceptional miners around making $200 an hour or better who were available for a single woman to choose from. Professionally speaking, I was dwarfed by those guys.

To my surprise Ana Maria did show at the agreed-upon time. We had a nice afternoon talking and a few laughs, but I ended up having to drop my alter ego, Rogelio.

After that day we saw a lot of each other, mostly on weekends for a couple of months, until one day Ana Maria announced she was moving to El Paso. The reasons for the move weren't clear to me or I wasn't hearing well, but either way there wasn't anything I could do about it.

I was disappointed but took solace in that she encouraged me to visit her in El Paso. I looked forward to that, and soon after I began a long streak of visiting El Paso every weekend.

At the end of every Friday shift at Section 35, I'd be off on the three-hundred-mile drive armed with a case of Tab and my tape collection. It took about four hours to make the trip and, sometimes significantly less if the state police patrol officers weren't out and about.

Socorro, New Mexico, had a state police office in town, and those who knew the route were aware that speed traps were often set up among the gently rolling hills of I-25 just south of town. Of course there was a price to be paid get that kind of information, and being one of the speeders, I was caught once.

I would drive the route from Grants to El Paso at between eighty-five and ninety. The speed limit at the time was fifty-five. I came up over a hill just south of Socorro doing at least ninety, and there in the center of the interstate was a state police car. Doing ninety miles per hour in a fifty-five zone was going to get me pulled over, so as I approached the police unit, I quickly hit the brakes and was on the side of the road waiting before the officer even had time to turn onto the road.

When the officer approached my car, he said, "That was pretty good. Thanks for stopping so quickly. I'll tell you what I'll do. You're getting a ticket for doing sixty-five instead of ninety-two. It'll save you some money."

The officer did write me the ticket and then said, "Slow down. Got it?"

"Yup, I got it. Thanks."

Ever grateful for such leniency, I of course had no intention of slowing down, except through Socorro where to this day I always lift off the gas pedal a little.

For the most part, those trips to El Paso were immensely enjoyable. The summer evenings were warm, and the breezes always seemed to be up, and the company of Ana Maria was wonderful. The night life was a respite from the weekly grind at Kerr-McGee and of a kind entirely unavailable in Grants. Still, with visions of Marty Robbins's "El Paso" rattling around in my head, I'm sure the reality was far less exotic than it felt to me at the time.

I liked El Paso and, having visited many times since in subsequent years, still do. While El Paso was great, Ciudad Juarez was better.

The northern Mexican city of Ciudad Juarez lies just south of El Paso on the other side of the Rio Grande River. Taken together the two cities have a combined population of about two and a half million.

Spanish explorers in the 1650s founded Juarez. Named for the exiled Mexican president Benito Juarez, it began as a city along the El Camino Real de Tierra Adentro (the Royal Road of the Interior Land) as part of a northern trade route through the southern Rockies all the way to Santa Fe. Many sections of the original trail, in spots running parallel to I-25, are still clearly visible. Evidently Juarez was as much a magnificent place to rest and recuperate in the 1650s as it was in the 1970s.

Ana Maria and I would frequently drive across the Bridge of the Americas and into Juarez to visit bars and restaurants, shop for groceries, go to the outdoor mall, or sometimes just to drive the city streets.

Getting into Mexico was as easy as it got. Mexican customs agents never once had us pull over for any reason. I usually slowed down just enough to see the hand signals the agents would give to either keep on going or to stop.

Our favorite night spot was the Camino Real Hotel. A beautiful old hotel, the Camino Real featured mariachi bands in the downstairs bar and disco upstairs. I greatly preferred the downstairs because of the local flavor it afforded.

It was there in the mariachi bar that I was introduced to real nachos, not the Doritos with Cheez Whiz, but the good stuff with handmade tortilla chips, refried beans, melted cheese, and some very hot jalapeno pepper slices. These the hotel staff would serve on gigantic platters. While I wasn't much of a beer drinker, I found the local beer to go very well with those nachos.

It was rare that I knew what the mariachis were singing about, but it was beautiful, relaxing music. Sometimes we'd go out by the pool under the palms and listen to the music wafting through the open windows of the mariachi bar and just talk and talk all night until it was time to head back to the States.

Getting back into the United States was at first always interesting. It was unusual for me not to be pulled over on the U.S. side of the bridge by U.S. Customs and Border Protection agents. I don't know what profile it was that the customs agents were working off of, but I know we, or I, matched it.

Often the agents would search only the car, but occasionally they would take dogs through the vehicle. That worried me because I imagined the possibility someone had planted drugs in the hubcaps or somewhere else under the car, planning to follow us and pick up their stash when we stopped. It never happened but seemed a reasonable possibility at the time.

Often during a thorough search, customs agents directed us to stand behind large Plexiglas or maybe bullet-resistant screens. Then we learned how not to get stopped.

U.S. agents always seemed to ask the same questions, revolving around how long we had been in Mexico and what we'd been doing. When an agent asked, "How long have you been in Mexico?" the incorrect answers included "Not long" and "Just doing some shopping." Those answers got us pulled over every time.

The bakeries in Mexico seemed to always serve only warm goods. The fresh tortillas were incredible, as were the many different types of rolls. They were worth a special trip. I rarely missed an opportunity to try a new bakery or revisit an old one. I once went over the bridge for some pastries at a magnificent bakery there in Juarez. I picked up the pastries, turned around, and went back to El Paso.

When customs agents asked how long I had been in Mexico, I said, "About twenty minutes." That triggered a full inspection by agents and dogs and found me standing behind the bulletproof barrier.

Getting tired of being pulled over, I tried experimenting with different answers until I found the one that always worked. The correct answer to the question of how long had I been in Mexico was "I don't know. We've been partying over at the Camino Real," with an emphasis on partying and acting just slightly inebriated. That answer worked every time, effectively ending the stops and searches.

Even after my weekly forays into Mexico ended, I continued to use that partying story technique, and it worked every time. The older I got, the less I needed to use it until finally I had only to act my age to get over the border. These days I mostly get bored looks from U.S. Customs and Border Protection agents.

When we weren't in Juarez, Ana Maria and I would be exploring the night spots and restaurants in El Paso, of which there were many excellent ones. El Paso by the Rio Grande is indeed a good city, and well worth a song or two.

Sadly, my fabulous weekends in El Paso and Juarez came to an end when Ana Maria moved on to greener pastures.

Seeing her had been a wonderful break from the all-male world of the underground that I was living in. For my part I appreciated our time together and what seemed to me to be exotic adventures.

I was bothered by that breakup for quite some time—a little too long, as it so happened.

Fuzzmobile

Al Friedt was yet another recent Grants High School graduate. He had grown up in Grants, and I think there were others in his family working in the mines.

Al was full of enthusiasm of a kind I hadn't seen underground. His aura was quite apart from the more basic motives of most miners. The good miners I knew went about their work driven by the need to earn a living first, then they got involved in competing with other miners to move up on the contract board, and then they became motivated by pride in their work. Al's enthusiasm was different.

Al was a high-energy guy, a highly enthusiastic, happy young man who had no idea where he was going but was going to get there in a hurry. Maybe it was a consequence of youthful enthusiasm, but I think there was more to it than that. He epitomized Cal's dual axioms of work like a mule and do something even if it's wrong.

In lieu of any better idea, Al headed for the mines after graduation. Not being sure he wanted to work underground, he had started on the surface driving a front-end loader. Loaders had very large buckets and huge rubber tires and were used to load ore trucks, move sand to the sand-fill hopper, and clean up around the ore hoist under the headframe.

A lot of guys I worked with didn't have any good stories of their time at Section 35 or didn't know how to tell them, but Al already had some front-end loader misadventures to talk about.

Ideally front-end loader drivers would not drive up the sand pile but rather take sand gradually from the bottom. Doing it that way was easy, smart, safe, efficient, but evidently not much fun.

For some reason Al had decided one day to take his loader up the side of a huge sand pile. Apparently he was trying to be the pioneer of off-road front-end loader driving and wanted to test the limits of the machine. There was no other work-related explanation for trying that. But a big pile of sand was not the most stable terrain on which to drive a loader.

Loaders had large enclosed cabs with doors on either side that opened outward. Should a loader start to tip for any reason, it was easy for the operator to escape if need be by jumping out the door opposite of the direction the machine was tipping, thus avoiding being crushed.

As Al made his way up the sand pile, his loader started to slide sideways. Seeing that the whole machine was going to flip, Al jumped. He jumped out the wrong door, however, in the direction the machine was going to roll. Had the loader rolled over onto him, he would have been crushed, but he was somehow able to scramble away just in time, and it missed him. Both he and the loader continued rolling down the side of the mountain of sand.

It came to be a legendary story among the surface crew, but at the time of Al's assignment to me I had heard nothing of it.

After that episode, Shotgun or Mel decided a better place for Al might be underground, and Al agreed, so he was transferred to me. Bill was the one who told me Al would be my new helper. Since I knew Al hadn't been working underground, I asked around and heard some rumors about a loader incident. I heard it was bad but didn't get many details other than Al was pretty stupid.

That worried me, because if he made a big mistake underground, we might not escape. Not having a choice regarding my new helper, I told Al it was great having him, but there was something I needed to know first.

"Tell me about the incident with the loader," I said.

I was expecting a lot of excuses, denials, and misdirected blame, but what I got was a very funny story and the truth. I quickly learned that Al was far from stupid.

Satisfied and impressed, I came clean about the light in the stope. If it was going to be a problem, I needed to know right then.

"We have kind of an odd light that keeps showing up in the stope that you need to know about," I explained. "Nothing happens and there's nobody there, but it keeps showing up, so if you have a problem with that, let Bill know now. If not let's go."

Al's response: "Oh yeah? What kind of light? I gotta see this. Let's go."

He seemed amused by my explanation of the mysterious light. I could tell he wasn't buying the story, but I had the feeling if and when he saw the light it wouldn't matter, so I felt much better about getting to work that day.

When we reached the stope, I still had drilling to do, so I picked that up while Al gathered supplies and stacked timber.

It was midmorning when the light appeared. I stopped drilling long enough to point it out to Al, but he had already seen it.

"Hey, there's a light; somebody's coming," he said.

"No, that's the light I was telling you about."

"Nah, it's too bright. That's somebody."

"It isn't somebody, Al; it's just a light that keeps showing up, but if you want to go see for yourself, go ahead."

Just like all of us had done previously, Al took off to see who it was, and just like everyone else, he was back shortly, having been unable to see or find anyone.

"Does that thing show up all the time?" he asked.

"Nope, just a couple of times a shift at the most, and it never moves around."

At that point he agreed that nobody was there. Unlike me, he thought the stope probably had a ghost, which seemed to amuse him greatly. He told me the ghost didn't matter to him as long as it stayed put.

For the three months or so that Al and I worked together, we saw the light many times. It usually never moved but now and then would shift a little from side to side.

At some point we noticed that the light was no longer making its daily appearance. It was gone, and we never saw the light again.

The relationship I had with Al was much different than Cal and I had. Al was about five years my junior, which made a difference, so we had some banter that went back and forth, unlike with Cal. On the other hand, much like Cal, he didn't find the humor in many of the things that happened underground the way that I did.

In fact, Al had just a bit of a sense of humor deficit. His serious side would come out from time to time in conversations as time went by.

While most miners would spice up every sentence with some profane pronouncement or other, I was more refined. Where many would say, "Fuck, that boss is a real asshole," or "Shit, that fucking mucker never works," I would say, "Boy, that boss is a real asshole," or "Boy, that fucking mucker never works." It was second nature to me and in some way seemed to add a more civilized tone to life underground.

Al and I had been working together a month when one afternoon, we had completed hoisting some very heavy twelve-foot timbers, and we were carrying them one by one back to the working area of the stope. I said, "Boy, these fuckers are heavy."

Al dropped his end. As I was on the other end, it hurt when that happened because of the added weight, so I dropped mine too. He looked at me as serious as could be and said, "Stop calling me boy. You've been calling me boy every day for a month, and I don't like it."

I started laughing and couldn't stop while Al stood there with a look on his face that said he was ready to do battle. Fortunately, I was so much bigger than Al that he didn't attack me, although I'm sure he wanted to.

After I calmed down, I explained that "boy" was just an expression I added to most sentences. It took me a while, but I convinced him it wasn't personal. His outburst persuaded me not to add "boy" to my sentences any longer, and since that day in the stope, it is a rare occurrence when I do.

One day I told Al how glad I was that the day was almost over. Belying his nineteen years, he told me we would never have that day back, so, no, it was not a good thing the day was over. That was the Al Friedt credo and how he lived and worked.

Al worked so hard there were times I stood watching and laughing inside as he literally ran from place to place in the stope. I thought Al was the kind of guy that would make a great ballroom miner. He moved so fast I don't know that a slab would ever be able to catch him. The speed at which Al worked combined with his enthusiasm made it appear to me that he was headed for the stratosphere of the mining elite one day, and I doubted he would be with me long.

Still being on the twelve-hour shift and seeing as we got along so well together, Al and I decided we would try carpooling. So Al who lived in Grants, wouldn't have to drive all the way out to San Rafael, I

Underground and Radioactive

drove into Milan on the days Al drove and parked my El Camino in an empty lot.

It was quite a sight when, on the first day of Al's week to drive, he drove up in his 1970 Plymouth Barracuda. It was a slant-back, dark-green machine with some distinctive features.

The rear end had been lifted to an almost ridiculous level, and to such an extent that in order to stay seated on the inclined passenger seat, I sometimes had to brace myself with my hands on the dashboard. The nose of the car was pointed into the ground, and just getting into the car was an ordeal. I learned to open the door and hold it with my right hand; then, putting my left hand on the dashboard, I slid in and pulled the door closed all in one motion. It was the only car I have ever had to fight my way inside.

Once in and semi-seated, I looked around the interior and noticed that everything in the car was covered with a fuzzy fabric of some kind. It was on the seats, the dashboard, the instrument cluster, the rearview mirror, the steering wheel, the rear seating area, the arm rests, and almost every other part of the passenger compartment. He topped it off with fuzzy dice hanging from the rearview mirror. Considering Al's humor allotment, I might have been taking a chance, but without thinking I blurted out, "It's a Fuzzmobile. Have you thought about fuzzwall tires?" From then on it was the Fuzzmobile.

The Fuzzmobile had seat belts, so although we seemed to be going down the road in a position somewhere between sitting and standing, it was possible to stay fastened into the passenger seat.

I had a lot of fun riding in the Fuzzmobile, and no matter how much I kidded Al about the car, he never objected. If I pointed out a spot in the car that he might have missed fuzzing over, he would have some fuzz on it the next time I saw it. He must have had quite a supply of fuzzy material at home. I never did convince him to go with the fuzzwalls, though.

Even with Al, who I liked, I never could get comfortable in car-pooling. So, despite my love of the Fuzzmobile and recognizing Al as being one of the safer drivers of my carpool experiences, I went back to driving myself to work.

The Fuzzmobile was not long for the commute out to Ambrosia Lake. Shortly after I stopped riding with Al, he either had an accident in it or decided to quit subjecting the car to the abuse of the long commute.

Al replaced the Fuzzmobile with one of the all-time classic vehicles in commuting history.

Al lived at home. I think his parents must have charged him quite a lot in rent because, although he was making good money for a nineteen-year-old, he always seemed to be broke. As a result, he replaced the relatively luxurious Fuzzmobile with one of the worst-looking, used-up, broken-down pickup trucks I'd ever seen.

At one time the truck had probably been a 1960s Ford, but it was hard to tell as all the identifying emblems were gone. Most of us at the mine agreed that it had the general shape of a Ford, but we had no idea what model it might have been.

Whatever interior insignia had come with the truck was long gone, along with the paint, bumpers, tailgate, and grille. Al covered the exterior with several cans, more or less, of black spray paint.

Though not exactly the chick magnet that the Fuzzmobile was, the truck did run and made the sixty-mile round trip to and from the mine every day with no problem, except for one.

The commute out thirty miles to Ambrosia Lake went by way of Route 53 in Milan to Route 334 to Route 509, all two-lane roads. What wasn't being mined was mostly open-range ranch land.

There is a long, flat sweeping curve along Route 334 that, if you were going slow enough to look, opens up into a beautiful panorama of the surrounding desert. Few miners were interested in savoring beautiful vistas while on the way to or returning from work. The speeds at which commuters traveled the long curve in the road put a good deal of centrifugal stress on both vehicle and occupants. I was always holding on to something in an attempt to stay seated through this curve during my time commuting with Boots, for example.

After my shift one afternoon, it so happened that Al was right behind me as we left work and drove down 509. He was still there when we made the turn onto 334. I would usually make the trip at a conservative eighty to eighty-five miles per hour. Al's truck could hardly make eighty-five miles per hour, but he had managed to keep up, and as we entered the long, sweeping curve, he was still behind me.

Halfway through the turn, I looked in my rearview mirror, and there I saw Al's right front wheel bouncing off down the road, over a small hill, and onto range land, where it continued in high arching bounces until it bounced out of sight.

To my amazement Al was still right behind me, cruising along at eighty miles an hour on three wheels. I was laughing so hard it was blind luck that I didn't run off the road. He never stopped. He had enough weight on the left side of the truck to keep it upright, so he kept on coming all the way around the curve but then had slowed considerably, and I lost sight of him. I couldn't let this one go, so I turned around and went back.

Al's truck, now off the road on the shoulder, rested on what was left of the right front axle. The axle had broken off. That will happen when a wheel turns on bearings that have been screaming for grease for a month. Al said he didn't know what the noise was, so he'd kept on driving.

Al was a level-headed, even-tempered guy who rarely got upset about anything. The twelve-hour shifts were exhausting week after week, but he never complained about it. As in any job, if you get tired, you make mistakes. Mining is one of the occupations in which mistakes can carry a high price tag, so we all tried to keep them to a minimum.

I was doing a good job looking after Al, but he worked so hard and fast it was difficult. Considering the combination of fatigue and the speed at which we worked, something was bound to happen, and it did.

Our stope was looking good. We continued to enlarge it until it was four stories of timber sets. We were filling the chute on a regular basis and making decent money. I always gave Al four hours of contract time, so he was happy, and I was happy with my thirty dollars an hour. All in all, we had a good working stope going.

Remembering how Cal rarely let me drill, I made it a point to teach Al as much as I knew about drilling, letting him drill rounds now and then. He picked it up quickly and loved it as much as I did.

One day we were working on the third level. We had blasted out the ore on the first two levels but hadn't mucked it because our chute was full, and as usual I was having a hard time getting a motorman to pull it.

We had a large muck pile going, having blasted out two stories' worth. In the meantime, we could still go up to the third level and keep drilling and blasting.

We were in need of a large number of supplies, but rather than have Al go back to the station, I decided to let him drill while I made the supply run. I figured I'd round up the stuff, find a motor, and deliver it to our stope myself. I could also use the then empty cars that had been full of supplies to pull some of the ore that was backed up in our chute.

We set up the drill, and Al started in on the round. Talking being impossible through the din of the machine, I signaled Al that I was leaving.

I made it back to the station but couldn't find Bill. Seeing a couple of laborers hanging around by the station, I gave them my list of supplies and told them to stack up the materiel and have it ready to go when I came back with a motor. It would take them a while to get everything ready, so I went back down the track drift to our stope to check on Al's drilling progress.

As I neared the manway, I heard the machine going. I knew Al was OK and still at it. I climbed the manway and then went up the ladder to the third level of square-sets.

As I looked toward where Al should have been drilling, I couldn't see him. I heard the drill going but didn't see Al where he was supposed to be.

I wasn't concerned until I got close enough to where he should have been, and still I saw nothing. But the drill was going, so he had to be there. Where was Al? What I found ranked right up there with the funniest things I would see underground.

I walked up to the edge of the set and saw the blasting holes he had drilled and, still hearing the drill going loud as ever, looked down.

Having somehow lost his balance while wrestling with the machine, he had fallen over the edge of the set and dropped down just a few feet. There below I saw Al on top of the muck pile on his back, jammed between the still-running machine and some previously blasted ore, feet sticking straight up in the air.

Al lucked out. Had we been mucking and filling the chute, it would have been a sixteen-foot drop, and he could have been seriously injured or worse. But he had fallen on top of the muck pile just a few feet below.

Had the machine fallen on him rather than beside him, that could have been serious too, but it had missed him. I jumped down, shut off the machine, and pulled him out, laughing the whole time.

Al and I did a lot of productive mining during the time we were together, and a lot of the credit for that went to him. I had never seen anyone work as hard or as fast or take his work more seriously. We had a good stope with good ore and a good system working, and we made a lot of money. We shipped a lot of ore, and the bosses were happy. It was too good to last, I guess.

Three months later Mel Vigil called me into his office at the end of my shift and asked if Al was ready to mine on his own. I didn't think I was the best judge of that, but seeing an opportunity to give a hard worker like Al a break, I said he was good to go.

Al was on his own.

Beginning of the End

Shortly after Al was promoted to miner, Mel told me I would no longer be working twelve-hour shifts. I had enjoyed the money, but the cumulative effect of working so many hours was exceptionally draining, both mentally and physically so, I was pleased to be assigned strictly day shifts and overjoyed at the prospect of seeing the sun every now and then.

Then Mel gave me some disturbing news.

"The geologists say there is still some high-grade stuff in the 502," he said. "What we want you to do is start digging the sand-fill out. Start at the manway, take the bulkhead out, and start digging. Go in straight about a hundred feet. Don't worry about left or right; just go in straight until you get to the face, and that's where you'll be drilling. It might take a while." I was stupefied.

This would be my third time working the 502 stope. First I was mining it with Cal when the stope collapsed. Then I was put in charge of sand-filling the place, and now I was to take the sand out and start mining it again. It sure made me wonder what was going on. There had to be some incredibly high-quality ore in there and a lot of it if the plan was to remove tons and tons of sand to get at it.

I had a lousy weekend after that talk with Mel.

When I arrived in the lunchroom on Monday, Bill had assigned me yet another partner, and it was my buddy from the stuck ore chute fiasco, Al Gonzalez.

The one tiny bit of good news was that we would be getting paid a small contract stipend for each ore car of sand-fill we removed. I think it was around ten dollars a carload, but at least it was something. It was definitely motivation to get a lot of sand out in a hurry.

The only way to get the sand out was to dig by hand using muck sticks, throwing the sand down the manway and then shoveling it onto the track, where we would load ore cars using a mucking machine. It was going to be a slow and laborious job.

During the sand-fill operation, I had constructed an eight-foot bulkhead to contain the sand as it was pumped into the stope. We removed the bulkhead, and there in front of us was an eight-foot high wall of sand. The wall was about ten feet wide at the bulkhead but quickly spread out as the stope widened going inward.

After staring at that impressive wall of sand for a few minutes, we slowly began tunneling into it. When we got around six feet in, things got a little dicey. Now we had a tunnel with eight-foot high walls of sand on either side of where we were working. Should either of those walls collapse while were we in the tunnel, we would be buried with no hope of getting out.

We decided that from then on only one of us would dig while the other stayed back where the bulkhead had been. Should a cave-in of sand occur, the one of us who was outside the tunnel could dig out the one who was buried.

It was very unpleasant work being inside that tunnel of sand with the eight-foot walls. Sand-fill carried with it a very unpleasant odor no matter how much ventilation bag we hung. In addition, the sand got into everything on our person like, but not exactly, a day at the beach. Then of course there was the possibility of a sand cave-in. When it was my turn to dig, I was somewhat nervous every minute I was in there.

The work was very slow. We could only throw so much sand down the manway before it would rise high enough to block our exiting the stope. We were constantly stopping the tunneling and descending the manway to shovel sand out onto the track.

When we had a good-sized pile of sand on the track, I would use a mucking machine to fill ore cars with sand. Running that mucking machine was the only fun part of the day for me. I loved running that mucker.

My previous experience operating a mucker—having worked for Riordan on track drift construction—came in handy, and I was very confident in my abilities. Although I derailed the machine every now and then, thanks to having so much practice returning derailed ore cars to the track, I was proficient at getting the mucker back on track.

Other than that mucker, what we were doing was a lousy job. There wasn't much about being underground that I disliked, but that job was one I detested.

End of the Line

Monday morning had arrived, and I was in a bad mood. I had been in El Paso again, for the thirteenth consecutive weekend, and it had not gone well.

That was the weekend Ana Maria and I had broken up with a fight starting just about the time I arrived late Friday evening. Rather than drive another four hours back to Grants, I made an ill-advised decision to stay in El Paso when I agreed to talk it over with her more on Saturday. Predictably that didn't go well either, so I was in a particularly bad mood on the long drive back to Grants.

I had few personal problems during my time working underground, something I didn't appreciate fully until years later. When there was something bothering me, I successfully avoided taking whatever it was to the job. Mining was too dangerous to be thinking about anything but the work.

Looking out for my partner and myself was always the main priority underground, and I had stuck to that rule. Yet here I was, distraught after having experienced a very bad weekend. I was functioning on four hours of sleep and not thinking much about the job.

My helper, Anthony Gonzales, and I had spent the past three months cleaning the sand-fill out of a very large portion of the 502. Clearing the sand-fill had been tedious, had not paid much, and was dangerous, but we had done it.

This day we were preparing to again mine the rich uranium that had been found by the geologists. We had set up our water and compressed air lines and had a nice new machine to drill with and a stack of timber waiting.

Before beginning the sand-fill operation in the 502 months earlier, I had removed all the valuable equipment, pipes, and hoses.

End of the Line

Among the items not removed were many scraps of wood left over from building the dozens of square-sets needed for ground support. I say scraps, but these ranged in size from small chunks to four-foot sections of twelve-by-twelve timber.

Because the sand sinks to the bottom and water stays on top, anything that floats ends up near the back during a sand-fill operation. The end result was a lot of large, waterlogged timber remnants pinned up to the back. That wouldn't have mattered, and nobody would have cared if the mined-out stope was sealed up. But in the case of the 502, Al and I had gone back in, making it a very dangerous place with all those timber remnants many feet above our heads.

As the sand was being cleared, we constantly kept an eye out for any timber scraps mixed in with the sand that could potentially fall on us. It wasn't unusual to have small portions of the walls of sand collapse on us as we dug. They were minor, but frequent. Having the walls of sand cave in on us was serious enough, but to get hit by one of those scraps could have been fatal, so we always stayed aware of where the waterlogged blocks of wood were. Similar to scaling the back and ribs after blasts, we often induced some sand cave-ins just to remove the danger posed by blocks of wood, taking great care when bringing them down.

We were now in the stope, having removed better than 95 percent of the sand-fill we needed to remove and could see the rich uranium ore we were after. This, I thought, would provide some very good paydays.

While the uranium ore was rich in the 502, the ground was no more stable than it had been during the days when Cal and I had mined there. In addition to watching for those chunks of waterlogged wood, we had to be vigilant for signs that the ground above might give way. In fact, having had many months of experience in there, I was much more apprehensive about the ground falling in on us than what condition the remaining sand-fill was in.

I had seen this place collapse before, and it didn't look much better or sound much different as Anthony and I prepared to drill and blast our first round in over three months.

As we got closer to the area where the ore was, I had spent some time surveying and planning. The face where we would be drilling didn't look good to me. The ground was very uneven and would need to be blasted flat in order to install proper ground support.

The immediate area where I would be drilling was an odd shape with many angles created by minor cave-ins and the overall of deterioration of the stope that had taken place prior to the sand-fill operation. All of the places I had mined previously had a square look to them similar to rooms and hallways. We built square-sets to hold the ground up, so we blasted square patterns to fit the sets into. Fairly simple. As I surveyed the area in front of us, I was thinking of ways to make it square and pretty.

I could see that the uneven ground had the black, sandy look of good uranium ore, but it wasn't flat, so wouldn't do. No, that place had to be made square, so I decided to make the floor nice and flat, and since it was good ore, we would get paid for doing it.

The area of the stope that we had cleared of sand was a fraction of the overall space that had previously been mined. We were to the left of the hundreds of collapsed square-sets from the previous cave-in that occurred when Cal and I worked the stope. All we had to our right was another eight-foot wall of sand approximately forty feet long.

It was usually prudent to plan wisely and work safely underground, but having been in a hurry, I unfortunately did neither in this case, and neglected to clear a larger area to our right that would have made the site where we would be drilling much safer. So there we were, preparing to drill just a few feet from a huge wall of unsupported sand. Furthermore, I should have considered that the reverberations caused by the drill could have potentially broken the wall loose, but it either never occurred to me, or I ignored the internal warning.

High in the upper right corner of our new stope face, I had seen a large piece of twelve-by-twelve square-set timber that I estimated to be three feet long. I knew from a lot of experience lifting similar remnants during the sand-fill clean-out phase that waterlogged pieces that size tended to be extremely heavy—so heavy, in fact, that it usually took both Anthony and me to lift one and move it to the side.

This particular piece was embedded in the sand at the corner of the exposed face on the right. It was clearly visible, and we should have brought it down before beginning to drill. While it was up high, the block of wood seemed to be far enough away from the small mound

End of the Line

that I intended to drill and blast flat that Anthony and I would be safe. It could be dealt with later when we actually got to drilling the face. Still, I had resolved to keep an eye on it and asked Al to do the same.

As was a helper's responsibility, Anthony was setting up the drilling machine while I surveyed the job, all the while obsessing over the trying weekend I had experienced. Having already planned out the operation in the days prior, I wasn't getting into much detail with myself about the job by considering options and alternatives. The end result was that I overlooked safety.

I have mentioned previously that when at full speed, the Ingersoll-Rand drilling machine made an ear-splitting racket and literally shook the entire environment in the immediate proximity of its operation. One downside was that talking or yelling to my partner became impossible with the machine running, so we relied on hand signals and stayed relatively close to one another for that purpose while the drill was on. Both partners constantly watched the back and ribs for signs of loose slabs and, when seeing one, would stop the drilling long enough to take care of it.

The vibrations the machine caused usually didn't have any effect on a work area properly supported. It was why most miners would never drill without stulls or a roof jack or both. Many, me included, would build square-sets right out to the face and use that for support and cover. But then, in the 502 things were different.

While this was not a ballroom, the back was too high in places to install a roof jack or a stull. The shape of the open area was too irregular for square-sets.

Anthony and I had used scaling bars and, in the spots we could reach, chipped away at the back, peeling off any piece of rock that had the potential for danger. Out in front of where I would be drilling, only that large chunk of wood high in the sand to the right posed any potential problem. If Anthony and I both kept an eye on it, I thought we would be fine.

With Anthony standing a few feet off to my left, I let the drill roar to life. Once again real production mining had returned to the 502.

With the machine going full speed, I stood there in the rejuvenated 502 with one hand on the throttle and let my mind wander back to the weekend. Most other days I would have been more aware of what was happening in the stope and might have seen the wall of sand that had been vibrated loose and was cascading toward me. But not that day.

The huge wall of sand to my right broke free along with that three-foot section of square-set post. Somehow I was not buried by the sand but was hit squarely in the back by the waterlogged chunk.

Getting hit hurt, and I knew in an instant that even if I didn't get buried by the sand, I wouldn't be getting up and walking out of this one. It was no little slab from which I stood up wondering what had happened. Oh no, this was something much more.

The rush of sand stopped, and I was left lying on top of it, miraculously saved from burial within. I lay there in agonizing pain, having great difficulty breathing, gasping for air.

Anthony Gonzalez, while grazed by the wall of sand, had managed to escape the worst of it by dashing toward the manway. When the sand stopped falling in, he had come running over to where I lay. He might have wished he hadn't.

I grabbed his arm and put a death grip on it that had to have hurt, and I wouldn't let go. It wasn't that he didn't try to leave to find help, because he did, but I wasn't letting him go anywhere.

Several minutes passed, and I was still having some difficulty breathing and still squeezing Anthony's arm for all I was worth. Then, slowly, I began to recover my breath.

As I experienced intense pain, it dawned on me that assistance was needed. It was then that I loosened my grip on Anthony. When I did so he was gone in a flash, down the manway on the way for help.

Section 35 had a rescue team that was fast, professional and serious. The guys I had come into contact with who worked rescue seemed to rarely smile, perhaps because they had seen too much or perhaps because so many depended on them. I had seen this team in action and knew that once Anthony let them know where I was, it wouldn't be long until they reached me.

Because the 502 was a relatively short distance to the station, it must have taken Anthony no longer than five minutes to reach a shift boss or a telephone. In approximately ten minutes, the rescue team arrived. Not knowing which would be needed, they brought along a rescue board used to transport miners suspected of sustaining a back or neck injury and a basket used for those having suffered, for example, a broken or missing limb.

The man leading the rescue team was James Baca, one of the premier miners at Section 35. He did jobs nobody else could do or would do, and he volunteered for both. He was making around $300 an hour from what I saw on the miner's contract board, and while I didn't know exactly what he did to earn that kind of money, I knew he must have been very good at whatever it was.

I didn't know James Baca, so here we were meeting for the first time, with me lying there doing all I could not to scream out in agony and him trying to get me to the surface as quickly as possible. I was reassured when James told me it wouldn't be long before they got me out of there.

I was in a tough spot, knowing I couldn't get up and save myself, knowing I had to rely on others for that. Fortunately, I could move around a little, and everything on my body seemed to be working, but there was definitely something wrong with my back.

Having ascertained the type of injury they were dealing with, the rescue crew took great care in lifting me onto the stretcher board, strapping me in and tying me down so that even had I wanted to move, it became impossible. Four members of the crew then picked me up and headed off toward the only way down to track level, the manway.

When we reached the manway, there was some discussion among the rescue crew as to how exactly I was to be lowered. It was finally agreed to attach the stretcher to the winch used to lift supplies and slowly lower me that way. There wouldn't be anyone with me on this trip, and it would be a slow one.

Fortunately, the manway was only about twenty-five feet down to track level, so while it was a slow trip, it was also a short one.

A few more of the rescue crew were waiting at the bottom of the manway, and when I arrived they carefully removed me, and we all headed to the station at the quickest pace they could manage.

When we reached the station, the cage was waiting, and up we went. I estimate that from the time of the accident to reaching the surface could not have been much more than twenty minutes, if that. I have always been grateful to the rescue personnel, themselves miners, who acted with such urgency and took such great care to get me out of there.

Now I was lying there in the Dry staring up at the curious faces. Thankfully Shotgun had everyone get back to what they were doing, and I was lying there alone, waiting for the ambulance.

Before long the ambulance arrived, and I was loaded up and ready for the very long ride to the Grants Clinic.

Aftermath

The diagnosis was a broken vertebra. I spent two weeks at the Grants Clinic, and while it wasn't pleasant overall, there were some enjoyable elements. One was Demerol.

At first the injections were urgently anticipated to ease the pain, and later I wanted one whether I needed it or not. My physician, Dr. Valdivia, was good and realized what patients like me were up to, so the administration of Demerol by needle tapered off and was replaced with a pill form of the drug that, while decent, couldn't match the rush of the injections. I now have at least some rudimentary understanding of drug addiction. That Demerol was good stuff.

I signed some papers and cashed some checks during the period of Demerol injections that I later had no recollection of. I ended up getting copies of the checks I signed before I would believe I had actually done it. That's what drugs will do to you.

As for the injury itself, I never lost the use of any extremity, but I did lose feeling in several spots on my body that ranged from a square inch to one area of the chest that was several inches around. I would pinch these areas and feel nothing for many months afterward.

Although my injury did not feel minor, during my stay at the clinic I met other miners with injuries far more severe than mine. I quickly realized the good fortune with which I had been blessed to have had such relatively modest damage.

The Grants Clinic. that also served as the only hospital in town, was filled with underground workers who had been hurt in some particularly gruesome ways. One of my roommates had been routinely operating a slusher when the drum bearings came apart, sending a bolt through his ankle, irreparably destroying his bones.

Another laborer had been walking down a main track drift when

Aftermath

he was struck by the lead ore car of a motor. He told me he never saw or heard it coming. He lived to talk about it, but he was pretty broken up. My discomfort paled by comparison to these men, and I resolved to tone down my internal self-pity a little.

My hospital stay lasted for two weeks, after which I was fitted with a waist-to-neck body cast and sent home.

At some point during my stay at the Grants Clinic, Greg Hornaday left without a word. I never figured that one out. He just up and vanished, never to be seen again.

Someone stole the battery out of my El Camino during my clinic stay, adding insult to injury, so I had to get that replaced just to get home.

The cast stayed on for three months, during which time it was somewhat difficult to get around or do much of anything. Occasionally I would take a drive, but had I suffered a flat tire, I would have been unable to change it, so I never ventured far from home.

During those months I watched a lot of TV and went for some walks in the San Rafael area, but that was about the extent of it for quite some time. When walking around got old, I went to a western outfitter in town and found a shirt that was two or three sizes too big but would fit over my body cast. Doing so I was able to go to the Iron Blossom by wearing my huge western shirt over the cast. I talked myself into thinking it was fine, but I can only imagine how preposterous that must have looked. Good thing we were all drinking.

Those few trips didn't work out well, and it wasn't the same as when I had been working. It was as if I wasn't a part of the underground fraternity anymore, so after a couple of attempts at it, I quit going.

I had ample opportunity to reflect during that three-month period, so that's what I did. Although I loved working underground, I had to admit it wasn't a real safe environment. Besides my own, I'd seen a lot of injuries and many far worse.

I really wasn't the best miner, or close to it, and wasn't sure I ever would be or ever wanted to be for that matter. Mostly I just liked the people I worked with, the honest hard work I did, the money and being underground.

The time was rapidly approaching when I would have to decide to either go back out to Section 35 or quit. Quitting would be no hardship, as I had managed to save a good chunk of what I had made.

Aftermath

I was medically cleared to return to work on a Friday yet still hadn't decided what do. After further considering everything over the weekend, I made myself a lunch, filled up my thermos, and headed back out to Section 35, still unsure whether I was going to work or quit.

I had with me a well-worn, beaten, grayish looking hard hat, thoroughly broken in overalls and my beat up boots. Although I'd been gone for a few months any new guys out by the headframe, who didn't know who I was, would assuredly know *what* I was. No greenhorn I.

But, somewhere along the thirty-mile stretch of highway out to Ambrosia Lake, it came to me that this was it. I was done with mining.

Pulling into the parking lot for what was to be the final time, I was already turning melancholy, but my mind was made up.

Entering the Dry, I was greeted by several of my fellow miners welcoming me back, but in reality I was already gone, and in passing didn't have much to say to them, as I was no longer a part of the subterranean fraternity.

Making my way down the hallway of the mine offices, I stopped and knocked on Shotgun's door. He was seated behind his desk looking over some paperwork and when he looked up at me quietly standing there I said, "I'm quitting."

Thankfully, doubtless with my wellbeing in mind, he did not attempt to talk me out it because he would likely have succeeded in doing so. Instead, "OK. Sorry to see you go, but good luck," was all he had to say.

My mining days now over I left the Dry feeling a little down, but glancing to my right was immediately buoyed by the sight of a small group of new hands, they of the shiny hard hats and clean clothes, bunched together near the headframe. Smiling to myself I thought, Good luck, I hope you have as much fun as I did.

Epilogue

I never saw Cal Cargill again, but I kept track of him as best I could over the years. He returned to his life and family in Montana. He passed away in 2006, so I suppose he wasn't as old as I thought he was back in 1976.

Shotgun Buchanan worked for many years as a mining engineer in Montana, New Mexico, Nevada, Oregon, Arizona, and even Indonesia. He eventually returned home to Whitehall, Montana where he passed away in 2016.

After Greg Hornaday vanished, I never saw him again but later learned that he continued on in underground mining and, later, construction work as an explosives expert. He must have been good to have lasted so long.

Al Friedt, the hard-working and good-natured helper, I rarely saw at Section 35 after we were no longer partners. I knew from looking at the contract postings that he did well. I heard he hung on mining uranium until the mines shut down in the 1980s.

Anthony Gonzales I never saw or heard from again, but I do hope he managed to get his driver's license and his Firebird back. Oh, and sorry about that death grip on your arm, and thanks for getting help, Anthony.

My first boss underground, Frankie Garcia, I never heard a word about until many, many years later, when I discovered that my Albuquerque barber was his brother-in-law. Small world. I wonder if he remembers the knucklehead laborer he sent to open an ore chute?

Bill Clark and many others from the Butte, Montana, area either returned to Butte after the uranium boom ended in Grants or fanned out to areas of the country where underground miners were still in demand.

Epilogue

The author in 2016 (photograph by Marti Martienssen).

As it was such a fascinating life to have lived in a boomtown and to have worked in a uranium mine alongside such an extraordinary group of people, I'm fairly certain, had I not been injured, I would have continued mining for quite some time.

I stayed in New Mexico, continued on with my education, and had a decent enough career in economic research. Not many days go by that I don't miss being underground though.

As eventually befalls most mining boomtowns, when the ore runs out or the demand for the end product disappears, Grants is a ghost of its former self, having been hard hit by the downturn in uranium prices in the early 1980s, due primarily to imported ore and the cessation of nuclear power plant construction.

Now, with the mines closed and the thousands of miners and their families gone, there are a few boarded up businesses and relatively few

Epilogue

vehicles on the streets of Grants. Even so, every time I visit, what I envision driving down First Street or Santa Fe Avenue, is a vibrant community, bumper-to-bumper traffic, the long defunct Iron Blossom full of boisterous miners and a line of vehicles waiting to fill up at the now abandoned Kerr-McGee service station.

Although my former home nearby is long gone, San Rafael appears today as it did during the uranium boom; a beautiful green oasis.

A massive cleanup effort has helped to remediate some of the damage that was done as a result of the mining and milling process, and is ongoing. Most of the land has the look of stunning high desert country once again. It's difficult to tell from afar there were ever any mines in the area at all.

On the long empty stretches of two lane blacktop, it is a leisurely, scenic drive out to Ambrosia Lake these days. Gone are the thousands of mine employees racing to and from work; the hundreds upon hundreds of vehicles having been replaced only by an occasional traveler now.

The former entranceway to Kermac's Section 35, and the rutted road beyond remain, but the rusted gate is now closed.

Mining Terminology

Back—The ceiling or roof of an underground opening.

Backfill—Waste material used to fill the void created by mining an ore body.

Block caving—An inexpensive method of mining in which large blocks of ore are undercut, causing the ore to break or cave under its own weight.

Cage—The conveyance used to transport men and equipment between the surface and the mine levels.

Chute—An opening, usually constructed of timber and equipped with a gate, through which ore is drawn from a stope into mine cars.

Development—Underground work carried out for the purpose of opening up a mineral deposit. Includes shaft sinking, crosscutting, drifting, and raising.

Drift—A horizontal underground opening that follows along the length of a vein or rock formation, as opposed to a crosscut, which crosses the rock formation.

Dry—A building where an underground worker changes into working clothes.

Face—The end of a drift, crosscut, or stope in which work is taking place.

Grizzly (or mantle)—A grating, usually constructed of steel rails, placed over the top of a chute or ore pass for the purpose of stopping large pieces of rock or ore that might hang up in the pass.

Hoist—The machine used for raising and lowering the cage or other conveyance in a shaft.

Jackleg—A percussion drill used for drifting or stoping that is mounted on a telescopic leg, which has an extension of about 2.5 meters. The leg and machine are hinged so that the drill need not be in the same direction as the leg.

Mining Terminology

Lagging—Planks of unfinished lumber that have many uses in a mine.

Manway—An entry used exclusively for personnel to travel from the shaft bottom or drift mouth to the working section; it is always on the intake air side in gassy mines. Also, a small passage at one side or both sides of a breast. It is used as a traveling way for the miner and sometimes as an airway, or chute, or both.

Misfire—The complete or partial failure of a blasting charge to explode as planned.

Raise—Similar to a manway but used mostly for lifting material into a stope.

Rock bolting—The act of supporting openings in rock with steel bolts anchored in holes drilled especially for this purpose.

Skip—A self-dumping bucket used in a shaft for hoisting ore or rock.

Slag—The vitreous mass separated from the fused metals in the smelting process.

Slash—The process of blasting rock from the side of an underground opening to widen the opening.

Station—An enlargement of a shaft made for storing and handling equipment and for driving drifts at that elevation.

Stope—An excavation from which ore has been removed in a series of steps. Usually applied to highly inclined or vertical veins but frequently used as a synonym for room and pillar mining.

Tailings—Material rejected from a mill after most of the recoverable valuable minerals have been extracted.

Tailings pond—A low-lying depression used to confine tailings, the prime function of which is to allow enough time for heavy metals to settle out or for cyanide to be destroyed before water is discharged into the local watershed.

Index

Numbers in ***bold italics*** refer to pages with photographs.

Acoma Pueblo 4, 34
Ambrosia Lake 1, ***3***–4, 26–28, 30, 41, ***45***, 47, ***49***, 79, 112, 124, 128, 158, 186–187, 203, 206
Anaconda 26

back *see* mining terminology
ballroom 37, 39, 141, 155–157, 185, 197
blasting board 52, ***143***, 145, 154
Bloomington Country Club 14–15
Buchanan, Arnold *see* Shotgun
Bustos, Manuel 94–98, 104

cage *see* mining terminology
call bell ***50***
Cargill, Calvin 116–117, 119, 123–138, 204
Carter, Bob 17–24
chute *see* mining terminology
Clark, Bill 112, 140, 204
Coal Mine Campground 32, 34–35, 81

drift *see* mining terminology
dry *see* mining terminology

El Malpais 35

face *see* mining terminology
Fort Wingate ***87***–88
Friedt, Al 182–190, 204

Garcia, Frankie 44, 52, 55–56, 62, 64–70, 72, 74–77, 94, 104–105, 109, 134, 204

Gardner-Denver ***149***
Gonzalez, Al 104, 191
Gonzalez, Anthony 68–76, 194–198, 204
Grants Clinic 30–32, 200–202
Grants Mineral Belt 1
grants 4–5, 24–26, 29–31, 33–35, 82–83, 87–88, 99–101, 122, 138, 175–179
grizzly *see* mining terminology
Gulick Hall 11

headframe 26–28, 47, ***49***, 133, 182, 203; *see also* hoist
Higgins, Tom 17, 20
hoist *see* mining terminology
hoist operator 48, ***50***
Homestake Mining 27
Hornaday, Greg 24–25, 28–29, 33, 35, 46, 81–83, 85–86, 88, 175–177, 202

Illinois State University 14, 17, 21
Illinois Wesleyan University 13.24, ***9***; *see also* Gulick Hall; Magill Hall
Ingersoll-Rand 148–***149***, 197
Iron Blossom 99–101, 202, 206

jackleg *see* mining terminology
James, Tom 171–173

Kermac 24–25, 27, 30–31, 36–39, 43, 78, 84, 86, 103, 125, 128, 130, 159, 163, 171, 206
Kerr-McGee 25, 37, 39–40, 175–176, 179, 206; *see also* Kermac

Index

La Ventana 99, 101–102
lagging *see* mining terminology
Laguna Pueblo 4, 34

Magill Hall 7, 13–14
manway *see* mining terminology
Martinez, Art 142–145
Milan 4, 34–35, 87–88, 138, 175, 186–187
mining terminology 207, 208
misfire *see* mining terminology
Mitchell, Gary 7, 8, 24, 120–122
motor 55–57, 59–60
motormen 48, 57, 59–61, 63, 69, 95–97, 155
Mount Taylor 34–35, **45**, 88
mucker 89–90, 92, 164, 192–193

Navajo Reservation 4, 34
New Mexico Mining Museum 4–5, **42**, **50–52**, **58**, **60**, **63**, **76**, **90–91**, **114**, **118**, **143**, **149**, **151**, **153**, **162**
nitroglycerine 68, 104–105

ore car 56–57, 60, **63**–64, 69, **76**, 78, 89–90, 92, 96–98, 165–166, 191–193, 202
Ortiz, Daniel 113, 168–171

Peters, Jim 106–111
Phillips Petroleum 26
pillar stope 94–**95**, **140**–141, 150, 169; *see also* mining terminology
powder 68–73, 75, 77, 79, 103–107, 147–148, 150–**151**, 153–157
powder box 68, 103–105
probes 136, 171–**172**

raise *see* mining terminology
Ranchers Explorations 26
Randolph, Boots 133, 187
rescue board **52**, 198
Riordan, Al 77–79, 89, 94, 103–104, 151, 192
rock bolt *see* mining terminology

rock drill *see* jackleg
roof Jack *see* mining terminology

San Rafael **87**–88, 112, 185, 202, 206
Sanchez, Jerry 96–97
sand-fill 112–116, 142, 144, 146, 182, 191–192, 194–196
Schultz, Oscar 77, 79
self-rescue unit **42**, 46, 108
Shotgun 6, 24–25, 28–30, 40, 44, 109, 118–119, 128, 137, 158, 168, 170–171, 173–174, 183, 200, 203–204
skip *see* mining terminology
slusher 116, 152–**153**, 201
slusher bucket 117–**118**
square-set **126**–131, 140, 146–147, 152, 157–159, 161, 165, 169, 173, 189, 195–197
State Farm 16–19
station *see* mining terminology
stope *see also* ballroom; mining terminology; pillar stope; timber stope
stull *see* mining terminology
Stutts, Jack 159–160, 163, 165–171

tamping stick **151**, 157
timber stope 127, 129–130, 140–141, 145, 150, 152, 155, 159, 169; *see also* mining terminology
trains *see* motors
Trang, Nick 14–15
Trang, Terry 14–15
trip light *see* mining terminology

United Western Minerals 26

ventilation tube **65**, 78
Vigil, Mel 28–30, 44, 75, 109, 137, 139, 145, 158, 171, 183, 190–191, 195

Wackenhut Security 16–18, 24
Western Nuclear 26

Zuni Pueblo 4, 34

www.ingramcontent.com/pod-product-compliance
Ingram Content Group UK Ltd.
Pitfield, Milton Keynes, MK11 3LW, UK
UKHW042000140426
5217IPUK00015B/901